SOLE
SURVIVOR

Daniel Rubin

Best wishes

Daniel

First published by Canbury Press 2025
This edition published 2025
Publisher: Canbury Press (www.canburypress.com)
14 Beresford Rd, London, KT2 6LR, United Kingdom
EU Authorised Representative: Easy Access System Europe
- Mustamäe tee 50, 10621 Tallinn, Estonia, gpsr.requests@easproject.com

ISBNs:
9781914487866 (hardback)
9781914487873 (epub)

SOLE
SURVIVOR

Daniel Rubin

For Anne
and Edward and Olivia

Contents

Introduction

I am a shoe obsessive. I look at people's feet before their faces; I can usually tell which brand they are wearing. My phone is full of thousands of photos of shoes taken from shop windows over the years.

I own over 100 pairs of shoes. I love the product. It is architectural and complex. It takes more than 150 operations, some requiring great skill, to make a pair of shoes. Unlike other items of clothing, shoes are stressed, pounded, and subjected to daily wear and tear. The full weight of the human body is applied daily to the sole and, in particular, to the heel of the shoe. The pressure on the tip of a woman's stiletto heel is reputedly equivalent to more than 15 times the pressure on an elephant's foot. For this reason the selection of the components and production of the shoe must be totally accurate, or the shoe will fail in wear.

Handling a pair of leather shoes is an experience that can't be replicated by seeing it online. The feel and smell of the leather, the detail of the stitching, the shape of the toe, the balance of the heel make it a special item of clothing. The styles range wildly, from flat ballerinas to over-the-knee boots, from functional sports footwear to highly embellished occasion wear sandals with 10cm heels. This variety makes the product both fascinating and challenging.

Shoes have a mythical quality. Cinderella's glass slippers were a symbol of enchantment and beauty (as well as showing the importance of fit). Dorothy's red shoes in the *Wizard of Oz* had magical qualities. They helped her to triumph over evil forces and liberate the oppressed people of Oz. I am not sure shoes can cause a revolution, but they can certainly transform your mood as well as make a statement about who you are. Whether it is winkle pickers of the 1960s, high platform sandals of the 1970s or red soled Christian Louboutin heels and Nike Air Jordans popular today, what you wear on your feet tells you a lot about your personality and sense of style.

The footwear industry is a big one: $400bn worth of shoes were sold in 2023; 24 billion pairs of shoes were produced. Over the years the price of shoes has relentlessly fallen as more and more cheap shoes are produced. Shoes have become a disposable item as the fast fashion trend has accelerated. Investing in a quality pair of leather shoes that lasts is one of the simplest and best solutions to meet this challenge and help solve the urgent problem of the 22 billion pairs of shoes that go into landfill each year. That's a big environmental footprint.

Having worked in shoes for the past 50 years, I am an industry veteran, a shoe dog. It's in my blood. I come from a long line of shoe manufacturers so devoting most of my working life to footwear is no surprise. For the past three decades I have grown Dune into a global footwear and accessories brand with a strong retail presence. In this book I look back on my journey in this period of immense change in all aspects of the footwear industry, which closely mirrors the changes that have taken place generally in the UK in manufacturing and retail. Manufacturing has largely left the UK and gone to China and low-cost countries in the East, while the emergence of the internet and changes in consumers' spending habits have substantially changed how we shop.

When I started in the industry in 1976 London had many factories making shoes. Factories could also be found in Leicester, Northampton, Somerset (the home of Clarks), Norwich, Rossendale Valley in Lancashire and a few in Wales, Scotland and Northern Ireland. Now, only Northampton has survived as a major area, because of the quality and craftsmanship of its factories. Most of the cheaper footwear is now produced in China and the Far East, where 87 per cent of the world's shoes are made, although India, eastern and southern Europe and Brazil are big producers. Italy still turns out the best quality shoes, having retained the artisanship and expertise to make beautiful shoes. It has the very best components suppliers and tanneries which, combined with generations of families who have been steeped in the art of crafting shoes, make the product so special and impossible to imitate.

Perhaps the biggest change in footwear over the past 20 years has been the growth of trainers or sneakers. Walk down any high street or shopping area and most people are wearing Adidas, Nike or another sports brand. Trainers started as performance footwear for running or sport but as health, fitness and in particular comfort have risen in importance, so have they. I sometimes wonder if, in 1976, I had gone into trainers rather than fashion shoes, life would have been very different. While Phil Knight, founder of Nike, was sourcing his trainers from Japan and South Korea in the early 1980s I was travelling to Taiwan to find fashion shoes. Mind you, Phil Knight was a talented American athlete from Oregon selling into the biggest consumer market in the world, and I was an Englishman from London with a family history of making women's fashion shoes, operating in a market a tenth of the size.

The change in the way shoes are sold has been as dramatic as the way they are manufactured. Clothing and department stores, discounters and supermarkets have found they can sell shoes from modest space in their stores. Without the need for expensive dedicated shops they can sell shoes at lower prices than specialist footwear retailers. They are just a commodity like their other merchandise. This has led to the demise of the mid-to-lower market specialists, with only a handful still surviving.

The digital age has dramatically changed shopping habits, with online shopping now accounting for 30 per cent of retail sales. These days if you have stores and are not an omnichannel retailer with a seamless offer to customers that includes in-store and online you will struggle. It is no coincidence that in 1998, when the British Shoe Corporation (once the biggest footwear retailer in the world with 2,500 stores) collapsed, the online behemoth, Amazon, launched in the UK.

For decades, I have been battling to keep up with and to get ahead of these changes. With hindsight I was slow to react at times. I kept my factory in London for five years too long amid the imminent collapse of manufacturing in the UK. Labour rates in the Far East were much lower and the workers were skilled and hard-working. I continued importing and wholesaling to multiple retailers when markets opened up and they were going direct to suppliers. Then I saw the light and started my own brand in 1992: Dune. With stores around the world and wide distribution through our partners in over 30 countries, Dune has established itself as a leading affordable luxury footwear and accessories brand. There have been ups and downs, and moments of high drama, when I feared I was going to lose everything. Fortunately there have been a few more ups than downs.

My father's repeated advice was: don't go into footwear; it's a tough industry. I didn't follow his advice. I sometimes wish I had.

1. Origins

Shoes are in my family. I am at least the fourth generation of the Rubin family in the footwear industry. My grandfather, Morris Rubin, was struggling to survive in Vilnius in Russia (now the capital of Lithuania) in the 1890s. Life in Russia at the end of the 19th century was very difficult, especially for Jews. The economy had collapsed and anti-Semitism was rife. Following the assassination of Alexander II of Russia in 1881 and the accession to the throne of his son Alexander III, who reversed most of the liberal policies of his father and was antagonistic towards the Jews, the Russians started the pogroms – a series of violent attacks on Jews. They were held responsible for the emperor's death, the dire state of the economy and anything else that was going wrong. Their homes were looted, their villages pillaged, and their businesses destroyed. Many were killed. This resulted in the biggest wave of Jewish emigration from Russia and Eastern Europe to the West in history. In the UK alone over 200,000 Jewish immigrants arrived between 1870 and 1910

Morris's elder brother, Monish, had emigrated to New York and Morris, who was in his late teens, decided to join him. Morris scraped together the 100 roubles (about six months' wages) to pay the fixer who arranged the voyage to America. He first had to make the arduous two-day journey from Vilnius to Memel (now called Klaipeda), the main port on the Baltic Sea. When he got to Memel, he was quickly herded to a small airless cabin in the hold of the boat where all those who didn't have a passport were placed. The unscrupulous smugglers had bribed the port and customs officials to allow those passengers without proper documents onto the boat. The voyage was horrendous. The cabin was overcrowded and filthy. Morris and his fellow passengers were not allowed out of the cabin. The food was practically inedible. They were like prisoners in a particularly unpleasant cell. The journey was made more perilous and difficult by the rough conditions in the North Sea. A few of the passengers, unused to being on a boat, suffered from sea sickness, adding to the terrible stench.

The boat finally docked at a port and the captain ushered Morris and others off the vessel. He was relieved to breathe fresh air. But Monish was nowhere to be seen. Like quite a few immigrants, Morris had been tricked. Instead of New York he had been offloaded in London. All he had in his pocket was one German mark. At the dockside, fraudsters, who spoke Yiddish, a mixture of German and Hebrew, the language of the immigrants, offered the confused arrivals accommodation at boarding houses. They were tricked into parting with most of their money and possessions. Morris, who spoke only Yiddish, remained at the dockside and slept rough for a few days.

Fortunately, he had the address of a distant friend of the family who owned a small shoe workshop in London. After a few sleepless nights, Morris, a cobbler, made his way to this friend's workshop in Whitechapel to ask for a job. He was in luck. He got a job as a cleaner and general helper. After a year he progressed to making shoes. He stayed in the workshop for a few years, working hard and sleeping on his workbench at night.

At the start of the 20th century there were a lot of Jewish immigrants in the East End of London living in harsh conditions.

Many families lived in one room, in the slums around Whitechapel and Spitalfields, where they also did outwork, such as sewing clothing and shoe uppers, for the local sweatshops. Although the conditions were harsh, they were a lot better than the penury and persecution they had endured in Russia.

Eventually things improved for Morris. In 1895 he married a determined and forceful woman from Vilnius called Rachel who was six years older than him. Rachel was ambitious and persuaded Morris to leave the workshop and start his own business making slippers. By 1911, they had seven children: four girls and three boys. The two eldest daughters worked with Morris and Rachel making the slippers, in their small apartment in Spitalfields. My father, Louis, the youngest child, was born in 1909. Two of his sisters died in childhood, a common occurrence in the years before the First World War in 1914. He had two elder brothers, Nathan (Nat) who was ten years older and Soloman (Solly) who was six years older. He was closest to his elder sister, Minnie.

In 1925, Morris bought a property on Vyner Street, Bethnal Green, in the East End of London, where he started his first manufacturing business, M Rubin & Sons Limited, with his sons, Nat and Solly, making women's shoes. He then moved to a larger factory on Mare Street, Hackney, where the business stayed until 1980.

Morris was an imposing looking man with broad shoulders and strong features.. He was quiet and reserved. Rachel was the dominant person in the Rubin household and Morris seemed to be happy to let her make the key decisions while he got on with working in the factory making shoes. My father, Louis, was musical and didn't want to work in the factory. He played the violin in an orchestra that performed during silent movies. Until 1927, all movies were silent as there was no way to synchronise the sound with the action on the screen. Louis was more cultured and ambitious than his brothers. He continued playing the violin until Morris (or more likely Rachel) told him this was no life for a young Jewish man. He should join the family making shoes. Reluctantly, that's what he did.

In the first half of the 20th century Britain had a substantial footwear industry (as late as 1956 it was the largest exporter of footwear in the world, exporting 12 million pairs of shoes). In the 1920s most major cities had shoe factories, as well as tanneries, component manufacturers and ancillary industries. In the retail trade there were a lot of small family-owned independents. Footwear styling was unexciting with women's fashion dominated by bar shoes on blocky heels and Oxford and Derby brogues for men and women.

M Rubin & Sons went from strength to strength as the demand for fashion and fashion footwear grew, especially after the Second World War. By the 1950s, it was one of about 50 shoe factories, many of them started by Jewish immigrants. Families like the Solomons, the Rosenblatts, the Wachmans and the Meltzers were all based in north-east London, manufacturing women's fashion shoes. The London manufacturers were renowned for making elegant and stylish footwear. Then, as now, footwear was a low-technology, high-labour business. Many operations were carried out by hand, starting with the clicking room where the leather upper was cut around the pattern with a special curved clicking knife. This was the most valuable operation in the factory as leather was the most expensive component of the shoe. For this reason, the clickers were the highest paid operatives. Their skill was in maximising the leather usage by placing the patterns in the optimal position so there was minimal waste. They also had to avoid any blemishes in the leather as these would ruin the shoes. Over time the cardboard patterns were replaced with metal cutting dies and hand cutting was replaced with clicking presses.

The biggest production unit in the factory was the closing room where the uppers were stitched together and attached to the lining. In many factories the stitching of the uppers was outsourced to outworkers paid on a piecework basis. The closing room was dominated by women on their sewing machines. There was a constant whirring noise as the Singer sewing machines stitched together the various parts of the upper. When the uppers were completed, they were passed to the lasting or making room. Here the upper was placed on a wooden last (which had an insole nailed to the bottom) to form the shoe. Lasting machines, that started to be used in America at the start of the 20th century, pulled the upper over the front and back of the last and attached them to the insole. The side of the shoe was pulled with pincers and attached with tacks. The sole and heel were attached, and the shoe was ready to be cleaned and packed. The clicking and lasting rooms were mainly staffed by male operatives, while the closing room and shoe room (where the shoes were inspected, cleaned and packed) were mainly female. M Rubin & Sons built up a reputation for making fine, elegant shoes. My father's brothers spent most of the day on the factory floor. Nat managed the lasting room and Solly the closing room. As demand grew so did the workforce. In the 1970s the factory employed about 100 employees.

My father was responsible for the finances, buying the materials, design and sales. He was a debonair man with a resemblance to the actor Cary Grant. He was a sharp dresser and very particular about his appearance. His suits were made at a Savile Row tailor, his shirts were handmade from a shop called Sulka in Jermyn Street in London, and his shoes were bespoke made from John Lobb, also in Jermyn Street, the primary road for high grade men's clothing and footwear. He went to the factory in a suit and tie. In the factory he replaced his jacket with a spotless white overall with his tie tucked into his shirt. In the factory he came across as a tough uncompromising leader who didn't suffer fools kindly. But his bark was louder than his bite. He was a sensitive and generous man and treated his staff like family. I heard lots of stories of him helping his employees financially and supporting them when they had difficulties.

On my rare visits to the factory in Mare Street, I was struck by the cacophony of sounds. From the lasting room came the clunking of the lasting machines, the hammering of the tacks, the screeching of the roughing machines, (which scoured the bottom of the leather uppers so that when the glue was applied, they adhered strongly to the sole) and the noise of the nails being driven into the heel by the attaching machine. The shoes were transported from operation to operation on racks which were on metal castors that made a shrill rattling noise. The heady odour of leather combined with the pungent smell of the adhesives that were pasted between the sole and the upper.

What I enjoyed most was the design room where the styles were sketched, and the pattern cutters created paper patterns. I was allowed to cut my own pattern which wasn't easy as it needs to be cut in such a way that it fits on the last. The last being curved and the paper pattern being flat, you needed to cover the last with masking tape, draw the pattern, and then place it on the paper to get the right shape. The head designer and pattern cutter was Fred Box. He was a true Cockney; a large generous man with a round ruddy face and thinning blond hair. He worked closely with my father and went on sales visits with him to customers, many of whom were based in Leicester. He was a warm-hearted, friendly man with a natural flair for designing shoes. He was devoted to Louis, whom he looked upon as a father figure.

Although there was competition between the factory owners there was also great camaraderie. Designers and pattern cutters moved from factory to factory. If a factory ran out of some component or leather their neighbour was very willing to let them have some of theirs. There was enough work to keep everyone on full production. Each factory had its own speciality. Meltzer was the largest and made more commercial volume styles. Rosenblatts specialised in plain court shoes. M Rubin & Sons had an elegant, more dressy aesthetic. In the 1950s and 60s, before imports from southern Europe and then the Far East started to pour into the country the manufacturers were in a strong position. They could be inflexible and limit their styling to their specialist production. This all changed with the arrival of imports. By the 1980s any manufacturer that survived had to adapt its production to whatever the market wanted.

The dominant footwear retailer was the British Shoe Corporation (BSC) which had been founded in 1956 by one of the prominent retail entrepreneurs of his day, Charles Clore. He had amalgamated several specialist footwear retailers such as Saxone, Dolcis and Freeman Hardy Willis. The BSC was one of the first retailers to embrace imports. Others soon followed. The benign conditions that British manufacturers experienced in the 1960s and 70s were about to change.

2. Early years

From the late 1950s, when I visited the factory in Mare Street, until 1975, I had very little to do with footwear apart from the occasional visit to the factory. My father didn't discuss his work much at home and I was too preoccupied with school and social life to raise the topic. He was a keen golfer and spent a lot of time at the golf club. I was aware that we were living a comfortable life in a large, detached house in a leafy suburb of London. We had enjoyable holidays driving through France (my father was a keen Francophile), taking holiday homes in Deauville in Normandy and staying at Relais & Châteaux hotels on the drive down to the Côte d'Azur. There were rumblings about the challenges of working with his brothers and family tensions but there was seemingly nothing out of the ordinary.

My parents had moved into a large, detached house in Hampstead Garden Suburb right after the war in 1945 which cost them £3,000. My father's income from the factory allowed them to live a very comfortable life.

My mother Dorie was a beautiful woman. She was petite with delicate features, bright blue eyes and a warm smile. She was utterly devoted to my father. She was a fantastic cook and kept an immaculate home. My father married her just after the war, when he was 34 and she was 17. My mother's family had a restaurant in Hackney, in the East End of London. My father met the then Dorie Goodman during the war when he was a special police constable. Footwear production was considered an important industry that had to keep operating, so he never went off to war. Louis would go to the restaurant for his dinner where my mother was the waitress.

I was close to my mother. I was much more comfortable and relaxed in her company. She was the one I went to when I had a problem. She helped me with my homework. I remember her trying to teach me maths, which was a struggle, as my attention span for the subject was limited. I was miles away, staring out the window, as she was explaining the intricacies of long division. I was a shy child and my relationship with my father was sometimes awkward. He couldn't understand why I avoided eye contact and had a weak handshake when I was introduced to his charity or golfing friends. He found that frustrating and disappointing as he was a sociable man who believed that it was important to give a strong first impression. Having said that, my two sisters and I had an enjoyable childhood and were brought up in a loving and generous home. It was clear that my parents were devoted to each other. We were very close to my mother's family. Friday night was spent visiting her parents, Jack and Bessie, where my aunts and uncles and their families crowded into their small living room for tea and strudel or cheesecake and had a natter about family matters, politics and the issues of the week.

Academically I was average. I always underperformed in exams as I got nervous and panicked. When I was 13, I took the Common Entrance Exam for St Paul's School, one of the top private secondary schools in the country. Competition to get a place, although tough, was a lot less than today and I was expected to get in. We got a letter from the Headmaster, Mr Gilkes, that said there was good news and bad news. The good news was that I had passed the examination. The bad news was that there was a Jewish quota and unfortunately, they couldn't offer me a place. I remember that my father was incensed and went to see Mr Gilkes to protest and try and persuade him that this wasn't reasonable but, although he was sympathetic and disagreed with the policy, there was nothing he could do to change the position.

The decision was made to send me to another public school, Stowe, in Buckingham as a boarder. No doubt my parents felt that I needed toughening up. I remember driving to Stowe with my father on the first day of term in September 1960. The building and grounds are grand and imposing and for a small and timid 13-year-old quite overwhelming. My father was driving a racing green vintage Mulliner Rolls-Royce which turned a few heads as he drove up and parked outside the front steps of this imposing Georgian building. He helped me schlep my trunk and tuck box out of the car, introduced me to my housemaster and turned to say goodbye. His parting words were: "Daniel, there are two things you should beware of: homosexuality and anti-Semitism." And with that he drove off.

To say that I didn't find it easy adjusting to life at Stowe is an understatement. Boys' boarding schools were tough places in the 1960s, even a more enlightened school like Stowe that prided itself on a liberal approach to education. I was desperately unhappy and cried most of the first term. It was all so foreign and different from the cossetted life I had lived up to then. Having to stand under a freezing cold shower for ten minutes, shivering with my teeth chattering, watched by a group of laughing boys as part of the informal induction process wasn't the best of welcomes.

Most of the boys were much bigger than me and had boarded at their primary so were used to living away from home. It took some time to acclimatise to this new life. I didn't do well in my A-levels the first time. There was a tuberculosis epidemic at the school, and I was diagnosed with a minor case of the disease just before my exams. I also wasn't very good at exams, didn't work very hard and was more interested in sport than studying. My tutor wrote: "The trouble is that he is such a slow worker that examinations easily expose his limitations. This is something that must be a handicap. The simplest way of removing it is not so much working under examination conditions as arranging his many ideas in his own mind."

I retook the A-levels and was offered a place at Warwick University to read History and Politics and Kent University to read Accounting. My father's consistent advice had been: "Don't become a footwear manufacturer – it's a tough life. Become a chartered accountant like Harvey." Harvey Cohen was my uncle, my mother's sister's husband. He was a successful accountant and financier and was admired by the family for his sharp mind and business acumen. Accounting, according to my father, was a great discipline for running a business.

It seemed strange that my father didn't want me to go into footwear, although it gave him (and us) a very comfortable life. He played golf on Wednesdays, Fridays and the weekends, paid for me and my two sisters to go to private schools, lived in a large, detached house in a nice area of London, had bespoke clothes, owned a Rolls-Royce and took us on expensive holidays. In those days you were more inclined to listen to your father's advice, especially if you had no clear vision as to what career to pursue. I had a vague idea that I wanted to own my own business. I am sure that was because most male members of my family had their own businesses, and the idea of being independent was appealing.

How had six years at Stowe prepared me for the future? It had certainly toughened me up. There was a lot about life at Stowe that I had enjoyed. The house, which used to be a large stately home, and the grounds (now owned by the National Trust) were magnificent. I was keen on sport and the facilities were excellent. I didn't regret not going to St Paul's. My parents were most probably right that I needed to get away from the comfortable life at home. It had instilled a strong sense of discipline and (something which is considered a bit old-fashioned today) good manners. It made me a lot more independent. It was a great education in how to live with and get on with lots of different types of people, which was valuable in later life. I made some good friends although after leaving school, unlike university, I had practically no contact with most of them. It wasn't easy integrating back into family life. My relationship with my father was slightly strained, which was more to do with my lack of social skills than anything else. Life at school and home were so different. During term time we only had three Saturdays a term to see our parents. So, when my father came to collect me in the summer of 1966 in his latest Rolls-Royce and helped me schlep my trunk into the car, I was a very different person to the one who had been deposited six years earlier.

I arrived at Kent University at Canterbury in September 1966 to read Accounting, an unusual choice given my interest in history and the arts subjects, but it was a choice that pleased my father. I had a vague notion that I wanted to go into business, be an entrepreneur, so an ability to read a set of accounts and understand the financial side of a business would be valuable even if accounting didn't sound like the most interesting subject to study at university. I had spent some time during the holidays at my uncle's accountancy practice.

As an intern the tasks I was given were undemanding and uninspiring, such as adding up the columns of telephone numbers from the telephone directory to improve my mental arithmetic. I wasn't very good at it. This reinforced my feeling that I didn't want a career in accountancy. I didn't have any clear idea as to what industry I would go into, except that I had been persuaded by my father that I should avoid footwear manufacturing, which was, with hindsight, sensible advice. The obvious move, after getting a degree, was to take articles with a large accounting firm and become a chartered accountant during which time I could decide on the future course of my career. Articles would take a further three years so there was no urgency in deciding my future.

The 1960s, the so-called Swinging Sixties, was a great time to be at university. Youth culture was a strong influence on society, and students were starting to have a voice. After the return to normality of the post-war years in the 1950s, the 1960s was a period of revolution. There were major changes in clothing, music and art. London became the centre of fashion with the King's Road and Carnaby Street in London setting the trend. The mini skirt, platform sandals and bell bottom trousers were replacing the more conservative fashions of the 1950s. Twiggy and Jean Shrimpton were the waif-like models who were the icons of swinging London. The Beatles were at the height of their power, following their hugely successful tour of the US in 1964, and they had a massive influence on fashion. Flower power and hippy culture became important trends culminating in the Woodstock festival in 1969.

Kent was not a hotbed of revolution, unlike the London School of Economics and other more radical institutions. This was partly because it was a new university, on a very pleasant campus, and still finding its feet and voice, and partly because the student population wasn't very political. I remember a headline in the student magazine, *Incant*, which read "Gilded Cage Tactics", which suggested that the students were living in this very comfortable bubble with quite restrictive rules and regulations with a limited interest in what was happening in the wider world. Kent certainly wasn't an environment that espoused radical causes although gradually the students got more involved in the broader student movement, joining events and marches organised by the National Union of Students to protest against authority, the restrictive rules in the university and supporting the wider call for the end of racial injustice and the Vietnam war.

In terms of dress, appearance and behaviour I also wasn't radical. I liked fashion and bought my clothes from the cool shops on Carnaby Street and King's Road, Lord John and Take Six. I grew my hair longish and had a beard for a couple of years. I tried cannabis a couple of times, but it made me feel sick, so I stopped. I was too scared to experiment with LSD or hard drugs. I wasn't aware of a lot of hard drugs being used in the university although experimental drug taking and "trips" on LSD were popular in the 1960s. I did smoke cigarettes, usually a brand called Embassy or Woodbines. I started my first cigarette in the evening and then proceeded to smoke pretty much non-stop until the pubs closed. On the more spiritual side I was persuaded to attend a séance but couldn't keep a straight face and was told to leave the room.

The only time that I got high on drugs was inadvertently at a party in Whitstable when the chocolate cake had been spiked with a large amount of cannabis. At the end of the party, I had agreed to drive a car full of friends back to the university campus at 3am. I was driving erratically, went through a couple of red lights, narrowly avoided a ditch and finally crashed the car on the roundabout near the university. Fortunately, no-one was hurt and there was a sense of relief that the only damage was to the axle of the car as it mounted the kerb. After that experience, I was wary of chocolate cakes.

The three years at Kent were very enjoyable. We were a privileged generation. Financially we were very well off compared to students today. We had no loans and received means-tested grants. Life was comfortable. In the first year and third years I had one of the study bedrooms in college and even had a lady who cleaned the room and did my washing. After the monastic life at Stowe the social life at Kent was hectic. Jazz in the bar area, dances in the main hall to the music of Tamla Motown, evenings at the Olive Branch in Canterbury and other local pubs, trips to Broadstairs, Whitstable and Herne Bay in the summer. There wasn't much time for studying.

In my first term I joined three friends in buying a clapped-out van for £30. Cars were not permitted but we felt that if we used the van surreptitiously to explore the surrounding area, including the resorts of Whitstable and Herne Bay where some of my friends were in digs, there was a minimal risk of being found out. It didn't work out like that. On our first evening of ownership, we were driving up the university approach road when the van came to an ominous halt. It wouldn't move. The only way it would go was in reverse gear. We were reversing erratically along the road, trying to keep the van in a straight line, when a tall, balding man waved us down. It was the Vice-Chancellor. Fortunately, he seemed more amused than annoyed. After that I bought a Honda 50 moped which was fine on a flat surface or going downhill, but going uphill, I was overtaken by pedestrians.

In July 1969 my parents came to the university to see me collect my degree. At that stage I was one of the few members of my family to go to university, so it was a special and proud occasion for them. I graduated with a 2.2 degree, worse than my tutors expected but a fair reflection on the effort I had put in. As well as being enjoyable, university improved my self-confidence. Having the ability at Kent, because it was small, to get involved and push myself to new challenges was an important lesson for later life. Although I was more confident, I still had a fear of rejection, especially in relationships. Maybe this was a hangover from my six years at Stowe. This remained during my twenties. I knew I had to push myself to take risks and not worry if I was turned down or it didn't work out, but that was easier said than done. I wasn't a natural extrovert or salesman but was gregarious and enjoyed mixing with different types of people. In that sense I was a bit of a chameleon. This ability to mix proved to be very valuable when I started my own business. Then I had no choice but to get out there and sell if I wanted to be successful. Worrying about rejection wasn't an option.

I reconnected with the university in 2017 when I was involved in setting up an extracurricular course called the Business Start-Up Journey to encourage entrepreneurs. In 2022 I was made a

Doctor of the University of Kent for my contribution to retail and my involvement in the university.

3. Being an accountant

Although I didn't approach the prospect with any great enthusiasm, having got a degree in Accounting, it made sense to join an accountancy firm and become a chartered accountant. At least it left my options open, assuming I survived three years as an articled clerk and passed the exams. In the late 1960s the demand for trainee accountants far outstripped the number of students that wanted to go into the profession, very different from today when competition is intense. Many of the big accounting firms now have an acceptance rate of around five per cent whereas in my day they were trying their best to persuade you to join their firm. It was another example of what a lucky generation we were compared with the youth of today.

I was invited to interview by four of the top firms in the City of London. I was offered a job by all of them except one called Arthur Andersen, which had a reputation for being more selective and targeted the most talented and ambitious students. Touche Ross was more relaxed. After a 20-minute interview, I was offered the job which I accepted. The offices were in London Wall in the City of London. Several of the more traditional staff still wore bowler hats with their three-piece suits although this practice was fortunately disappearing. I joined the army of formally dressed workers on the daily commute, squeezed into underground carriages and buses, to the City. It was a big change from the casual and informal life at university. There were no floral shirts or kipper ties here.

Most of the next four years was spent on audits and studying for my accountancy exams. The audits themselves were not particularly interesting. However, the opportunity to visit a wide variety of companies, get an insight into their culture, understand their finances and what made them tick was fascinating and educational. You also had an opportunity of meeting a lot of different people at varying levels of seniority. The auditors were not always welcomed with open arms. Generally, they were considered a necessary evil who distracted the team from their day work. I learned how to be diplomatic, ask the right questions, and not take up time unnecessarily.

My audits went from a small artificial flower maker in central London to Watney's brewery in Barnes to Barratt, which made sweets such as Refresher and Sherbet Fountain and had its factory in Wood Green in north London. I was also involved in the audit of the General Electric Company (GEC), one of the largest industrial and most profitable conglomerates in the UK. GEC was run by Lord Weinstock, known for his business acumen and financial caution. He was a serious-looking man with rimless glasses and the look of a courtroom judge. He was keen on controlling costs. In the toilets and other common areas there was a notice reminding staff to turn off the lights when they left to save electricity. It was signed by Lord Weinstock.

I audited a large cross-section of organisations including the Royal Society for the Blind, the Royal Mail Pension Scheme, several stockbrokers and Sea Containers, an interesting company that made a fortune in shipping containers round the world. James Sherwood, who built Sea Containers, was an American and a keen anglophile. He founded Sea Containers in 1965 with the head office in London, but the company was listed on the

New York stock exchange. He was a larger-than-life character, a serial entrepreneur, who in the end overextended himself. He bought the prestigious Hotel Cipriani in Venice because it was his favourite hotel, and then a whole chain of luxury hotels. I met him a few times and was impressed by his dynamism, energy and enthusiasm. He was a good advert for becoming an entrepreneur – a bon viveur and socialite, he made doing business fun. There was a big contrast between his risk-embracing, gung-ho approach to running a company and an auditor's conservative, cautious, risk-averse mindset. This was the main reason why I wasn't attracted to a career in accountancy.

Although I didn't have any natural aptitude for accountancy, I enjoyed the four years I spent at Touche Ross and, having passed my exams, had become a chartered accountant. I had learnt a lot, met some interesting and talented people and could now read a set of accounts and understand what made businesses tick. I was encouraged to stay on at Touche Ross but felt I needed a new challenge. Although my numerical skills had improved it wasn't a discipline I found easy or particularly enjoyed. One of my close friends at university was Chris Tanner. Like me he had joined Touche Ross where he remained for many years. He became a director and close adviser of my import and retail companies. Unlike me, he was an excellent accountant. He was good at the detail and seeing the downside of some of my more ill-conceived ideas. Over the years he was one of the first people I turned to for well considered and valuable advice.

In the early 1970s there was a lot of deal making in the City of London. A new breed of entrepreneurs was buying up industrial companies to create corporate conglomerates. The acquirers of these businesses weren't interested in growing these companies organically. They were making money by selling off the assets which were worth more than the company was capitalised on the stock market. They were called asset strippers. One of the most notorious of these was James Slater of Slater Walker Securities who built a sizable empire aggressively buying up undervalued companies. Many of these asset strippers branched out into secondary banking. Lending to businesses that often couldn't get credit from the main banks because they didn't have the collateral to support the loan. Slater Walker and many of these companies came to a sticky end as the economy worsened and credit became tight. Several had to be bailed out by the Bank of England and others went bust.

In 1974, I joined a new bank called London & Continental Bankers as a corporate finance executive. It was set up by Warburg's, one of the most prestigious and successful merchant banks at that time, and the shareholders were all cooperative banks in Europe. The aim was to support these banks and their clients with corporate finance expertise in areas such as mergers and acquisitions, risk capital and general corporate advice. It sounded like a good opportunity to get more experience in advising companies rather than just doing their audit. The only assignment I remember was a visit to the brewer, Kronenbourg, in Alsace in France. I don't remember much about the work, but I do remember I had an amazing *choucroute garnie* washed down with Kronenbourg 1664 beer and a spicy Gewürztraminer wine.

The work at the bank was uninspiring so I started looking for something more challenging. I read an advert in the *Financial Times* for a commercial finance director. An entrepreneur, Paul, had recently acquired a demolition company called Goodman Price which had been one of the main contractors involved in demolishing buildings in the City of London as part of its rapid redevelopment in the 1970s. The company, which was owned by the Price family, and which dated back to 1850, had been very profitable during this period. Coincidentally its head office was in Hackney, round the corner from where my grandparents owned a restaurant and where my mother had worked as a waitress and met my father during the war.

The plan was to use Goodman Price as a vehicle to buy into other successful trading companies. My role, apart from being a conventional finance director, was to review potential acquisition targets, investigate their finances and come up with a proposal that was compelling and made commercial sense. It sounded like a perfect opportunity to work closely with an expansion-minded entrepreneur and get involved in the more interesting aspects of finance. Unfortunately, it didn't work out as planned. Paul seemed to me to be more interested in buying the first Jaguar XJS sold in the UK and going on an extended holiday in Martinique than devoting his energies to building an industrial conglomerate. Meanwhile there were worrying legal problems with the acquisition of Goodman Price. The Price family had used the sale of the company to take money out of the company and this had fallen foul of the Companies Act which led to a prolonged legal battle. In the end the sale had to be reversed. I had received a valuable lesson in doing detailed due diligence before committing to a job or for that matter making any financial commitment.

My role changed from looking for new acquisitions to getting involved in the day-to-day activities of the company. Demolition companies make a proportion of the profit from the fees charged for the actual demolition work. However, a large profit could be made from salvaging and selling scrap materials from the demolished buildings. Goodman Price had great connections with architects and contractors in the City and a combination of getting some major lucrative contracts and astutely handling the sale of the scrap had resulted in the company making substantial profits. The challenge was to sustain this growth which, given the more difficult economic climate, was proving more difficult, especially with the distractions of the legal action.

A German company finally bought Goodman Price. It had developed a new, more efficient way of demolishing buildings, using special machinery rather than the more traditional method of demolishing by hand. It proposed that I become managing director, which was surprising and flattering, but I didn't see my future in demolition. It was 1976 and I was approaching my 30th birthday. I had to decide about the future.

4. Becoming a shoemaker

In the late 1960s my father had finally decided that working with his brothers at M Rubin & Sons was intolerable. There were frequent heated arguments which sometimes became physical. After a lot of angst and soul searching, he made the decision to leave. It wasn't an easy decision as it meant rupturing his relationship with his family, who from that day onwards we saw infrequently, only at family weddings and funerals. His father, Morris, had already become a more distant figure. It was a difficult period for my parents as leaving the business was full of financial risks as my father would leave with nothing, but this was preferable to working every day in such an unpleasant environment. My parents didn't talk about it much. I did get the sense that it was my mother's support and encouragement that finally persuaded my father to leave. She was the stabilising influence in their relationship, less emotional, more practical and decisive. I always felt that in a more modern age she would have been a successful administrator running a charity or a businesswoman, but she was happy being a supportive and dedicated housewife.

My father had had a health scare in 1955. He had been taken ill when we were on holiday in France and had to have a kidney removed. He appeared to make a full recovery and lived a full and normal life.

After leaving M Rubin & Sons, my father was introduced to a man called Jack Rose who had a factory in Stoke Newington, near Hackney, also making women's fashion shoes. Jack had made good money in the 1950s and 60s in the post-war years but the company was struggling and needed someone with my father's skills and more disciplined approach to revive it. My father bought a 50 per cent stake in Jack Rose Shoes Limited. Jack Rose was a character, a real rough diamond. He wore a camel coat, trilby hat, smoked a large cigar and drove a pink Roll-Royce. He had slicked back black hair, strong features and a confident expression that said: "Don't mess with me". He lived in a large house in one of the premier streets in Hampstead that he had painted pink but had to change to a more conventional colour after a complaint from the council.

What he lacked in finesse he made up in his energy and physical strength. Although he wasn't a tall man, he had the physique of a boxer. His handshake and back slapping left their mark. He was, however, given to bouts of depression and big mood swings. He was a generous man with a certain joie de vivre. His wife had died of cancer, and he was dating a lady called Joy. One day he arrived at her home in his Rolls-Royce, proposed to her, gave her a large diamond ring and told her to pack a bag as they were driving to the South of France to celebrate.

Despite their very different characters, my father and Jack seemed to get on well together. My father was cultured, thoughtful and well organised. I never heard my father swear. Jack was coarse, acted on instinct and took risks, not always ones that were well thought through. Jack notionally managed the workers in the factory while my father managed everything else.

During the fashion revolution of the 1960s and 70s the company traded well. They were making knee-high fashion boots for stores like Biba, one of the hot brands of the period. The boots complemented the miniskirts that were popular at the time. Sales were so strong that the factory couldn't meet the demand, despite working overtime. One of the many challenges of making footwear, as opposed to clothing, is that the production is limited to the capacity of the lasting machinery, unlike clothing where you can employ additional operatives to cut and stitch garments to boost production. This is one of the reasons why the lead times for footwear are longer and the stock held by footwear retailers is much larger. Daily deliveries were made to the iconic Biba store on Kensington Church Street, London, where customers queued for hours for a pair of boots.

In 1976, I was deciding what my next career move should be. Working my way up the corporate ladder did not appeal. Although I had enjoyed being an accountant and had learnt a lot, it wasn't a career I wanted to pursue. I was happy to work hard, had no commitments (but also no capital) and the prospect of owning and running my own business appealed. The question was in what industry should I devote my energies. A few of my friends had made a lot of money in the property market but that didn't appeal. I preferred a faster paced industry. I enjoyed cooking, going to restaurants and food generally so a business in this sector appealed. The British public were becoming increasingly interested in food. Celebrity chefs like Delia Smith inspired people to start cooking. Meanwhile more exotic foods were being imported into the UK and new restaurants were opening with varied cuisines. McDonald's had been an instant success on opening its first restaurant in the UK in 1974. That was its 3,000th restaurant worldwide so there were clearly opportunities of getting to scale if the concept was right. I was planning to go to America and look for a new restaurant idea that I could bring to the UK.

About this time my father's health deteriorated. He had a recurrence of his kidney problems. As I had a hiatus in my career, I decided to join him in the business to give him whatever support I could. I was reminded of his advice of not becoming a shoe manufacturer and to become a chartered accountant. I had become an accountant and the move into shoe manufacturing was only temporary – or so I thought.

Jack Rose Shoes was trading well. My father had found a niche making high-legged leather and suede boots as well as the more traditional fashion courts and sandals. He had been placing large contracts of leather at favourable prices. As the leather was the main component of the boot and he was able to cost at the current leather price, as opposed to the much lower price at which he had placed the contracts, the profit margins were high. He had an excellent relationship with the owners of the tanneries, and they respected his footwear-making skills and understanding of exactly the type of leather he required. He also had a natural charm that endeared him to most people he met. He was very particular about the qualities he was looking for in the leather. The hide had to be large to suit the size of a knee-length boot, with very few blemishes and with a stretch quality that suited the tight-fitting boots that were popular at the time. I remember him tugging at the leather to ensure it had the right amount of flex. He kept a detailed written record of all the leather contracts on a small notepad. He hadn't embraced technology and so had recorded everything by hand. I tried to persuade him to buy a photocopier to save on administrative work, but this was considered an unnecessary extravagance.

One of the main tanneries that supplied the leather was West Coast Tannery based in the Lake District in Cumbria. It was owned by a Hungarian immigrant who liked my father and enjoyed doing business with him. My father decided it would be good for my education to drive to the tannery to understand the tanning process. It is a messy and smelly experience, not for the squeamish. The process involved curing the skins with salt to prevent putrefaction. Soaking them to remove the salt. Soaking them again in a lime solution to remove the natural greases and fat. Removing the flesh by passing them through a machine and the hair by applying special chemicals. Treating them with salt and acid to prepare them for the tanning process and finally chrome tanning which involves placing them in a bath with chromium tanning agents which produce the wet blue hides as the hides at this stage are blue. There are other tanning processes, but chrome tanning is the most common and effective.

The leather was then cut to the correct thickness and finished to suit the end user's requirements. This is a simplification of what is a complicated, skilled and generally unattractive process. Leather is a by-product of the meat industry and therefore uses a material that would otherwise be wasted.

Joining Jack Rose Shoes was an eye-opening experience. I knew that making shoes was a complex process but was surprised at all the small operations that were required. Proper planning was essential as unless all the materials and components were ready at the right time production came to a halt. Late delivery of any of the components, however small, could hold up the whole production line. The uppers also had to be ready to be lasted. Some of the upper stitching was given to outworkers as the capacity in the factory wasn't sufficient. The production planner was constantly on the phone to the outworkers, who were mainly Greek Cypriot immigrants, to ensure that the uppers were delivered on time.

The factory itself was an old two-storey building which wasn't ideal for efficient production. The movement of uppers and components was done manually in boxes and on racks in the lasting room. One of the problems was that, unlike the factories in Brazil and the Far East, the production runs were short. As the factory was in the UK and lead times were fast the quantity ordered by our customers was often small. This meant that lasts had to be changed and machines reset, which adversely affected productivity. Getting big orders or repeat orders on styles was a bonus as the factory could operate without interruptions and changes, which optimised productivity. Keeping the factory full of orders was the key objective. Apart from the direct costs of the materials and labour an element of overhead recovery and profit was cost into each shoe. Therefore, the larger the production the higher the profit

During this period, I got much closer to my father. My relationship with him had been strained, largely because I had been away at school and then university since the age of 13 and found it difficult to adjust to being back at home. I know he was very proud that I had been to university, got a degree and qualified as an accountant. Working with him every day I got to appreciate that behind the confident exterior he was an emotional man who was a lot more vulnerable and caring than he appeared on the surface. The only real argument we had was about a girl I was seeing. I had met her in a nightclub in Sheffield on New Year's Eve while I was on the audit of the Post Office in Chesterfield. I was driving up to Sheffield most weekends to see her. I could see he was unhappy although he expressed this more by his facial expressions than mentioning it openly. Finally, he could hold back no longer and as we were driving to the factory one day he blurted out: "How long is this hanky panky going on?" He was keen that I meet a nice Jewish girl and get married, which I did, a few years later in 1977. One of my best decisions was marrying Anne Townsley. She has been my partner and soulmate for the past 48 years and a stabilising influence during my busy career in the shoe trade.

The first few weeks at Jack Rose Shoes were spent shadowing my father. His many years in the industry had given him a unique understanding and an ability to anticipate and resolve the many challenges that arose. He had a confidence borne out of experience. You could tell that his team had a lot of respect for him and his leadership skills. Although he was demanding and didn't hold back in expressing his displeasure when mistakes were made, he was a supportive and fair boss. There was a constant flow of managers coming into his office asking for advice, getting orders signed and problems resolved. It was a very hands-on management style but one that suited the nature of the business and one that he was very comfortable with.

Our largest customer was the British Shoe Corporation (BSC) based in Braunstone, Leicester. We supplied the more premium end of the group which were brands like Saxone, Dolcis, Manfield and Roland Cartier. A trip to the BSC headquarters at Braunstone in Leicester was a stressful experience. Firstly, you had a long wait in reception where the commissionaire, Terry, booked you in. Eventually you were ushered to a sample room on the second floor where you unpacked the shoes you wanted to show. Finally, the buyer came to look at the shoes and if you were lucky placed an order. Sometimes the buyer didn't appear but sent their apologies through one of their assistants. This procedure could take several hours and was a huge waste of time although there was good camaraderie between the different suppliers while we waited patiently to get a room. It was at the BSC I saw my first mobile phone in 1982. One of the suppliers was Alfred Magnus, a serious-looking man who had the appearance more of an accountant or lawyer than a shoe importer. He was a slight man with an educated demeanour, who always dressed in a suit and tie. He was carrying an enormous phone with him. It was called a Nokia Mobira Senator and was a large bit of equipment weighing around 10kg. It caused quite a stir.

There was a fast-track system for some of the big manufacturers from Brazil and Italy who went straight to the second floor and directly through to the buyers' offices. I never achieved that elevated status. However, often the orders they gave you made the wait worthwhile. I made that journey to the BSC for 20 years until the company finally closed.

Some talented buyers at BSC taught me a lot. The Saxone buyer, a lady called Cathy McKinnon, was a perfectionist. She gave you a small piece of material to indicate the colour she wanted for her shoes. It was very small, the size of a fingernail. You had to copy it and get her approval. This could take several visits as the shade had to be spot on, any slight variance was unacceptable. She also was very particular about the vamp height of a court shoe. That is the measurement from the toe to the opening of the shoe where the shoe meets the foot at the front, often called the throat. She measured the sample and wanted it to be exactly the right measurement which was often impossible as there is always a tolerance of two millimetres. This frustrating experience did teach me the importance of detail, and getting the dimensions as accurate as possible, because these make an important difference to the look and fit of the shoe.

Apart from the BSC the factory sold to specialist retailers like Barratts and Ravel, clothing retailers like River Island, who were becoming increasingly important as a destination for selling footwear and accessories, and the mail order companies like GUS, Freemans and Grattan, the precursors of today's online retailers. Mail order was big business partly because it offered attractive credit facilities, you paid in installments, and partly because it was often a social experience, getting together with friends to look through the catalogue. It was especially convenient for those who didn't have access to the shops.

My father had an interesting dispute with Barratts that was a valuable cautionary tale. Stylo Barratts was a large chain owned by the Ziff family based in Bradford. They were keen competitors of the BSC and at one stage if you supplied Barratts you couldn't supply the BSC. My father had sold Barratts a court shoe in several colours including green, which was very much a subsidiary colour that never sold well, unlike today where green has been a strong fashion colour. In those days the main colour palette was very predictable: black, beige and white in the summer and black, grey and burgundy in the winter. Barratts got delivery of the order and after a couple of weeks placed a large repeat on the green colourway. My father queried this with the buyer and asked him to check that this wasn't a mistake. "No," came back the response, there had been exceptional sales, and the order was correct. "Proceed with the order," which is what my father did. A few weeks later he received a telephone call cancelling the repeat order for the green shoes. What had happened was a troupe of Irish dancers had gone into the large Oxford Street, London store and bought all the green shoes for their show. Based on the sales, the buyer placed the repeat. He hadn't checked the system to see that Oxford Street was the only branch that had sold the green. The leather had already been placed, and the order couldn't be cancelled. Barratts insisted the order be cancelled. My father refused. The case was going to court until Barratts finally conceded and reluctantly accepted the order. I remember seeing a lot of green court shoes on their sale racks that season. Business with Barratts ended after that episode.

The London footwear manufacturers were a tight-knit community of companies founded largely, like my grandfather's, by eastern European and Russian Jewish immigrants and based in the Hackney, Dalston and Stoke Newington districts of north-east London. They were generally family-run businesses making between 5,000 and 15,000 pairs a week. The exception was Fiona Footwear which was founded and run by a charming man called Monty Sumray who began his career running a footwear trimmings and accessories business in London and ended up being the largest footwear manufacturing company in the UK making 170,000 pairs a week in his factory in Bridgend, Wales. Most of his production was sold to Marks & Spencer. He had the ambition, and management skills, to go for a large-scale manufacturing unit in the UK. Fiona Footwear was a model factory using some of the latest technology, although it too succumbed to competition from the Far East and closed in the early 1990s.

The manufacturers used to meet at charity dinners. I remember at one of these events a dapper man came up to me, shook my hand and said in a broad German accent: "I understand there is a history of insanity in your family." I was taken aback and must have looked confused. "Yes," he continued, "I understand that you are a chartered accountant and became a lady's shoe manufacturer. You must be insane." His name was Tony Franks, and he owned a factory in Hackney which many years later moved to Corby, a new town in Northamptonshire. There were times during my ten years as a manufacturer when I questioned my sanity spending so long trying to make a sustainable living out of making fashion shoes in London.

In September 1976 my father was taken seriously ill. He was rushed to hospital and after a few weeks he died. His other kidney had failed. He was 68. My mother was 51. The family was devastated.

In the short time that I worked with my father I learnt a lot and gained huge respect for his immense knowledge of the industry, his charisma and work ethic. He was held in high regard by his colleagues, suppliers, customers and competitors. After more than 40 years making shoes, he was still passionate about the product and kept a little sketchbook in his pocket where he did small drawings of shoes, often stopping someone in the street to sketch their shoes. One thing I inherited from him is that passion for shoes.

My father's death left a gaping hole in my mother's life. She never remarried and lived on her own. She couldn't conceive of being with anyone else. She was a fiercely independent and immensely capable lady. To keep her active, we opened a small shoe shop for her to run in Petticoat Lane, a market street in the East End of London. We underestimated the difficulties of owning one independent shoe shop, especially in an East End market, and closed it within a couple of years. My mother devoted the rest of her life to her family and doing charitable work.

5. Challenges of making shoes

One of the decisions we had to make as a family was whether to continue working with Jack Rose. The shareholders' agreement of the company contained an option that obliged the surviving shareholder, Jack, to buy out the shares of the deceased shareholder, my father, if the deceased's executors exercised the option. I liked Jack but his mental health had deteriorated. He became depressed and had started drinking heavily. I would see him standing by Andy, who operated the toe lasting machine, dressed in his suit, and every few minutes he would reach into the inside pocket of his jacket and pull out a silver flask and take a slug of whiskey. A few times he fell on the floor and had to be helped up and driven home. His drinking caused other problems. He was one of the cheque signatories and signed the cheques for the weekly wages. He had signed them all in pencil which had caused a panic in the office. It was sad to see him deteriorate but it became clear that I didn't want a partner who had a major drink problem. It was a shame as he had been very supportive after my father's death. Some of his involvement was less desirable. I was going to see the buyer of Freemans, the mail order company, when Jack stopped me as I was going out the door. He handed me an envelope and said, "Give this to John, the buyer. He's expecting it. Give it to him quietly when you both go to the toilet." He gave me a sly wink. Naively I did as he said. I must have been mad. That was the last envelope I handed anyone.

At the start of 1977 we sold our shares in Jack Rose Shoes Limited. We didn't get a fortune but enough for my mother to live a comfortable life. I was out of a job. This was a good opportunity to end my short career as a shoe manufacturer and look for an easier way of earning a living.

My uncle, Len Goodman (not the famous dancer), my mother's younger brother, had a company called London Lane Shoes Limited, also making women's fashion shoes. His company had been acquired by Clarks, then one of the largest manufacturers and retailers of footwear in the UK. The acquisition hadn't worked out and Len had bought back the company. The factory was in Dalston, just down the road from Jack Rose Shoes. At the time the company was not trading well. My family suggested that I join Len. After a few meetings we agreed that I buy a 50 per cent stake in the company. London Lane was missing a finance director so this was a role I could fill, as well as working with Len to improve profitability and grow the business. From memory my investment was £50,000 to buy half of the company. It was an easy decision. Here was a company run by someone I knew, admired and trusted. I could use my skill and experience to improve the performance, and I could buy into the company at a modest price. Jack Rose had traded well and made a reasonable profit. There was no reason we couldn't achieve the same success with London Lane. At that stage British manufacturing was doing well. It all made sense.

Len was honest and hard-working as well as being a sympathetic and supportive partner. He was a talented engineer and got great pleasure from working in the factory solving manufacturing problems or improving productivity. He was less strong on the commercial side of the business and often didn't have the patience to deal with some of the more demanding customers. He was well liked and respected by his team who had been with him for many years. As a young man he had spent several years in Israel working on a kibbutz and serving in the army. The ascetic life on the kibbutz suited his down-to-earth, forthright, equitable view of life. He was a good-looking man with wavy blond hair, blue eyes and an aquiline nose. His main downside was that he smoked 60 cigarettes a day. I shared an office with him so when I came home in the evening my clothes stank of smoke.

Len had a strong sense of justice. He didn't like people taking advantage of others. I went on a trip with him to a shoe and component exhibition in Paris. At the end of the day, we left the exhibition to get a taxi back to the hotel. We were met with a very long queue of people waiting for taxis (a frequent problem in Paris). We joined the back of the queue which was moving slowly. Everyone was getting frustrated. A group of men came out of the exhibition hall, went straight to the front of the queue and got in a taxi that had just arrived. Len had noticed this and didn't like it. He rushed up to the taxi as they were getting in and told them to get back in the queue. They ignored him and got in the taxi. This didn't please Len. He proceeded to stand in front of the taxi so that it couldn't move. The taxi driver could see that Len was not going to budge. Eventually he told the passengers to get out of the taxi, which they reluctantly did. The rest of the people queuing had seen this incident. When the offenders got out of the taxi and walked away a big cheer went up. As Len walked back to join us, other people in the queue patted him on the back and shook his hand.

London Lane Shoes was in an old two storey building in Downs Park Road, Dalston. There weren't any vintage fashion shops or hip cocktail bars in Dalston at that time. The crime rate was high. Crack cocaine dealers traded their wares on the corner of Dalston Junction. It wasn't an area where you popped out for a nice snack at lunch time.

The factory was not well suited to modern manufacturing. The clicking and closing rooms were on the first floor and the lasting and shoe rooms were on the ground floor. There was a small loading bay at the side of the building. Like Jack Rose Shoes the uppers and components were carried around in plastic boxes and the shoes transported in the lasting room on racks. The overall impression was of organised chaos. Getting all the uppers and multiple components ready for lasting required a major feat of planning, and the absence of any one component could bring the whole operation to a halt.

The job of organising the production was done by Derek Sewell, one of Len's key lieutenants, a quietly efficient and hard-working man in his fifties. He was tall with balding grey hair, blue eyes and a pointed nose. He was friendly, modest, and very good at his job. His role was crucial to the smooth running of the factory. I learnt a lot from Derek about the challenges of running a shoe factory and the importance of planning. He had a large board on his wall which mapped out the status of production of each order and where the various components relating to the order were. Today this planning would be done efficiently on a computer.

Len's other director was Tony Deakin, the factory manager. Tony was an ex-military man and had the demeanour of an army sergeant major, immaculately turned out with a starched shirt and tie. Managing the factory workers was not an easy job. There were several temperamental characters who were difficult to manage. There was a particularly fiery Greek Cypriot man on the toe lasting machine who had the habit of throwing the last across the room if the upper didn't fit well on the last. There were also outstanding managers like Breda, the Irish forelady of the shoe room (where the cleaning and boxing was done), who managed her department with an iron fist but was a warm hearted and generous lady with a wicked sense of humour.

Our biggest customer was the BSC, which Len managed. Most of the other customers I looked after. They included many of the same customers we had at Jack Rose Shoes; mail order companies (GUS, Freemans and Grattans), Ravel and River Island. River Island, which was then called Chelsea Girl, was one of the few clothing retailers to take footwear seriously. Footwear represented about 20 per cent of the company's turnover so was a major part of business. I dealt with two members of the Lewis family. First with Geoffrey who was the brother of the founder, Bernard. He was a colourful character. His office was like a French boudoir with a large painting of a woman's private parts on the wall. Then with Julian (Bernard's nephew), who was the exact opposite. He looked after IT and footwear, an unusual combination. He was a serious, bookish man, who clearly had a strong intellect but no apparent interest in fashion. Why he was given footwear was a mystery, but he was a pleasure to deal with as, like many good buyers, he listened to the supplier and was guided by us as to what styles to buy.

Inspiration for new designs came from lots of different sources: from the leading brands' catwalk shows, to what famous celebrities were wearing, to street fashion. One of the main sources for new ideas was visiting the trade shows in Milan and Dusseldorf. These two seasonal fairs were a great opportunity to get an understanding of the direction of fashion in footwear; what were the hot colours, toe shapes and profiles for the coming season. Were boots going to be strong? What height heels were in vogue? You could get a good overview of footwear trends by walking around these huge exhibition halls, with thousands of factories from all over the world, although you had to be careful that those trends would fit the UK market. Occasionally a trend would be big in Europe, but it didn't resonate with the UK customer. It was easy to get carried away with the exciting shoes you saw on a busy stand in Milan but, when you got home, in the cold light of day, they looked either quite ordinary or just not commercial, or, as we would say, "tricky".

As you walked down the aisles you would pass people you recognised and thought you knew, occasionally exchanging a friendly smile, but it was just that you had seen the same people walking around the shoe fairs over many years. They were most likely shoe dogs like me. The fairs were a great opportunity to network with suppliers and customers: to share ideas and discuss the many issues facing the industry.

One of these was the growing competition from manufacturers in the Far East. Each year several factories disappeared from the fair as competition became more and more intense and production moved from Europe to the Far East, particularly China. In recent years the fairs have got smaller with more and more empty stands. Even so, sometimes I would find a unique product that I would have missed if I hadn't attended the fair. Occasionally a version of this product turned out to be one of our best sellers, although finding these suppliers was often like searching for a needle in a haystack and required a large degree of luck. I would spend hours walking round the fairs looking for these hidden gems but eventually, suffering shoe blindness from seeing so many similar shoes, I would call it a day.

The Dusseldorf shoe fair was a big social gathering of UK buyers and suppliers. They met in the Irish pub in the old town and spent most of the night drinking. The next day you would see them wandering around the fair, bleary eyed, recovering from their hangover. Those days are long gone. Attendance at the fairs is well down and companies have set much stricter rules in employee behaviour. Visits to the fashion capitals of Paris and Milan were an opportunity to see the local shops to get inspiration and ideas for ranges. I must have taken thousands of photos over the years. It became almost an obsession. While walking round the shops during the shoe fairs you would see people in the industry from all over the world busy crowding round the shop windows to take their photographs. The attitude to taking photos of shop windows varied from country to country. In most countries the shopkeepers didn't see this as a problem. It was different in Paris where some of the shop owners were very protective of their shoes. They put their hands in front of the shoes so you couldn't photograph them or came out of their shop to confront you, gesticulating wildly. The secret was to photograph quickly and avoid eye contact. On one occasion the disagreement with an irate shop owner almost came to blows as he tried to snatch my camera. In Paris it was always easier to photograph at night or on a Sunday to avoid the hassle.

Back in Britain, the rise of cheaper alternative sources of shoes meant dealing with customers was an increasing challenge. Thankfully, the first season I joined London Lane Shoes we had great success with a wedge sandal called Daniella. The vamp was made from several leather strips in a crisscross pattern and the wedge was covered in raffia. We sold it to the Dolcis division of the BSC. It was a runaway hit. We received several repeats and new colours were added. Len built a special machine to make the strips for the vamp which greatly improved the production and gave us a competitive edge. It was a particularly warm summer, so Daniella continued to sell from April through to August.

The first three years went well. We had styles like Daniella which were best sellers. London Lane built a reputation of being a good quality supplier, which worked closely with its customers and designed fashionable styles that sold. The orders we received kept the factory full, which was the top priority. As a result, we were making a good profit. As we moved into the 1980s trade got more difficult as production moved overseas. Orders got smaller and were placed, later making the whole planning and production process much more challenging. Our prices started to become expensive as wages and overheads in the UK increased. Inflation was running at a massive 15 per cent which meant we had to increase prices to be profitable. Even with the falling pound the attraction of buying from southern Europe, Brazil and Asia grew as their prices were lower than the UK manufacturers. London had a particular challenge as wage rates were higher in the capital. As our prices were coming under pressure, we decided to make some shoes in synthetic materials to save on costs and make our prices more competitive. This was a mistake. We went back to making only leather shoes. Trying to compete with Asia on synthetic shoes was not a wise move.

We considered relocating the factory to Corby, a new town in the Midlands that gave attractive incentives to incoming companies and where labour and other costs were lower. Tony Franks (the man who had questioned my sanity) and his partner George Gives, had successfully moved their factory to Corby. In the end we decided against it as the challenge of transferring production and training a new workforce was fraught with difficulties.

Around this time, I got an interesting lesson in pricing from my cousin, Richard Rubin, my uncle Solly's son. Richard was a swarthy, good-looking man with a throaty voice who had a lot of charm. He mixed in a wealthy and fashionable crowd who frequented the "in" clubs and had extravagant parties. He would often jet down to the South of France to spend time on some debutante or peer of the realm's yacht or their villa in St Tropez. When the M Rubin & Sons factory closed, Richard started a range of women's shoes called John Smith, whose production he outsourced to London Lane. The shoes we made for him were very similar to ours except they were branded John Smith. What was annoying and revealing was that Richard managed to get a 20 per cent higher price than ours because of his persuasive branding and selling skills. This was a vital lesson in the power of a brand.

6. Importing from India

My first trip to India was with Len in 1980. We were making a cowboy boot in the factory and the cost of doing the decorative stitching on the vamp was prohibitively expensive in London. We found a factory in Agra, India, called Wasan, that would produce the leather and stitch the uppers at a competitive price. We arranged to visit the factory to ensure that the quality met our standards. Flying into Mumbai (then Bombay) over mile upon mile of shanty towns and then driving on a potholed road into the centre passing groups of young beggars tapping on our windows, a man defecating by the roadside, cows roaming around the streets was a salutary experience. Very different from modern Mumbai with its skyscrapers, motorways and sense of growth and dynamism.

We got the train to Agra and were met by one of the brothers, Deepak. The factory was old fashioned and extremely hot, but the quality of the uppers was excellent. It is always interesting when visiting a factory to see what they are making for other customers. At the same time, Wasan was making a moccasin for a Danish brand called Ecco which has been a hugely successful comfort shoe brand, with 2,250 stores in 101 countries. What was particularly impressive, apart from the unusual nature form design of the shoe, was that the owner's daughter, who was a young woman in her early twenties, was based in the factory for six months to ensure that the production of this Ecco moc met their quality standards. It takes that degree of dedication to build a successful global brand.

By 1984 the factory in Dalston was finding it increasingly difficult to maintain full production. The British Shoe Corporation and other customers were placing their orders later and later and the quantities were getting smaller. In the pecking order of priority, the Far East had become the number one source, followed by Brazil, then southern Europe, then the UK and finally the London factories. The reduced size and late placing of orders meant that production runs and lead times were shorter, putting a lot of pressure on the factory. The days of styles like Daniella, the wedge sandal that sold so well when I had joined in 1977, were long gone. These styles were now largely being placed in the Far East and Brazil.

We received the top-end styles that were not so commercially attractive to the customer because they were more expensive. The challenge faced by the factory was that these frequent changes lowered productivity. The changes disrupted the production flow as the new lasts meant resetting the lasting and other production machines. It also took time for the operatives to get used to the new constructions. Apart from the operational challenges the cost of opening a new set of lasts, knives and other components was expensive and these costs were now being amortised over a smaller order, which adversely affected the profit.

I decided to try my hand at importing. If we can't compete with the imports, let's try and join them. Len was preoccupied with the factory, and I found that I had some spare time. The finances, although not in rude health, given the size of the company and lack of complexity of the business model, were easy for me to manage. We set up a company called Capital

Shoes, initially to import sandals from India. The choice of India was a logical one as we had bought uppers from India and knew several sandal factories in Bombay.

We appointed an Indian agent called Atul Sanghani who had been working for one of the largest UK importers of Indian sandals. Atul was an articulate and experienced man who knew the Indian market well. He was soft-spoken and serious with a calm demeanour and an impassive face. He was plump, well dressed and had an impressive head of black hair. His role was not only to find factories but to ensure that the quality was good, and the orders were shipped on time. India produced handmade artisan sandals called chappals. These sandals were popular at the time. During the 1960s and 70s the Jesus sandal, or chappal, became the footwear of choice of the hippie movement and the trend had migrated into the 1980s. The Jesus sandal was typically a simple flat tan leather sandal with a few straps. Our aim was to create a range of more interesting sandals, flat and also on wedges.

We showed the small range to the BSC which placed a trial order. It was delivered on time and sold out in a few weeks. The BSC placed a repeat and added several new styles. The signs were encouraging. I flew to India to meet Atul and visit the factories in Bombay, Kanpur and Agra. We were met in Bombay by one of the factory owners who took us to dinner at the Taj Mahal hotel which looked like a large exotic palace with a big dome and intricate detail on the facade. I wanted a simple chicken curry. They chose the European restaurant on the top floor that was incredibly expensive. They insisted that I had the filet mignon that cost £200. They were an extremely wealthy family who wanted to impress. They owned the largest flour milling business in the region which evidently was a licence to print money. The son, who I had met in London, wanted to start manufacturing sandals. I should have advised him to stick to the flour milling.

The next day Atul and I went to the railway station to catch the train to Kanpur to meet the factory owners. The station was manic with huge numbers of people milling around. It was difficult to take in the sheer volume of people crowded into one space. The Asian Games were being held in India and all the flights to Kanpur were fully booked so we had to take the overnight train. Atul approached the train supervisor who said the train was very crowded. All the sleeping compartments were fully booked. If we wanted to travel on this train, we would have to sit in the unreserved general class, which was a closed-in cattle carriage with open slats to let the air in. I wasn't keen on spending the night squashed against other passengers in an enclosed carriage with no seating. Atul reassured me that we wouldn't have to travel in this carriage. Eventually after a lot of haggling, Atul gave the supervisor a few hundred rupees, and we were upgraded to sleeper class with two bunk beds. The train left the station with passengers jumping onto the train while it was leaving the station. Passengers were sitting on the roof and hanging out of the windows.

After a restless night we arrived at Kanpur at 5.30 in the morning. It was still dark with a heavy mist. The scene on the platform took me by surprise. It was full of people sleeping rough. Atul had disappeared for a few minutes to make the travel arrangements. We got our bags together and opened the carriage door. I was met with a long red carpet. Either side of the carpet were the eight factory owners (four each side). This was the welcoming party. They were smiling broadly and enthusiastically shaking my hand to welcome me to Kanpur. Half-awake I felt like some foreign dignitary arriving for an important conference rather than a novice shoe importer with a factory in Dalston. We were ushered into a car to begin the drive to the hotel to unpack and freshen up. As we were driving on the main highway into town there was a walkway across the main highway and draped across it was a large awning saying "Welcome Daniel Rubin of Capital Shoes of London". If the factories were as impressive as the arrival arrangements this was going to be a productive visit. We checked into the hotel which overlooked Green Park Test cricket stadium. I was shown to the room by the manager who said that this was the room that Ian Botham, the cricketing legend, had stayed a few weeks before during a recent test series.

After freshening up we started our tour of the factories. Shoe production in India was predominantly a hand-crafted business. None of the factories had shoe-making machinery. All the lasting was done by hand. The sandals were made in vegetable tanned leather, a process that goes back some 5,000 years when hides were soaked in water filled with Mimosa bark and leaves. The hides are soaked in several drums containing vegetable tanning agents for about 30 to 60 days including a drum containing tannin (a tree bark extract). The final leather, which has absorbed the tannins, is typically a natural earthy tan or brown colour with rich, deep patina. Vegetable tanned leather also has a sweet distinctive smell which is quite strong and not to everyone's liking. I could always tell which was my sample case of Indian sandals. As soon as I opened it, I was overwhelmed by the smell.

All the factories we visited were modest workshops with a few interconnected rooms where the different operations were carried out. The uppers, insoles and soles were cut and passed to the lasting room where the uppers were attached to the sole, often slotted through a slit in the leather insole. They were then stuck to the sole. The workers, mainly men, sat on the floor, cross-legged, making the sandals. They worked quickly and efficiently. The final product was sprayed and put on the roof to dry. It was a simple process that produced a distinctive type of ethnic sandal. France was one of the main markets for this type of natural sandal.

The factory owners were enthusiastic and excited by our partnership and the orders we were going to place. They knew the BSC and appreciated the volume orders they could receive if the styles performed well in store. We gave them a range of styles to develop which were more interesting and fashionable than the current Indian sandals on the market. Some were flat and others were on wooden wedges which they would shape in the factory. I left them feeling encouraged that this was a good opportunity to broaden our business and reduce our dependency on UK manufacturing. We visited factories in Agra and Bombay and decided where we would place the orders. I would return in a couple of months to check production with Atul. It was a quick trip and, like a lot of my overseas trips, I didn't have time to do any sightseeing.

India was a fascinating place. The profusion of people, noises, smells, colours and heat assaulted the senses. The UK seemed grey by comparison. After a couple of days I even got used to the poverty, which had shocked me at first. The pace of life was slow and there was little sense of urgency, which was frustrating, but despite the poverty most people were friendly, smiling, hospitable and keen to please. I had been told to be careful about what I ate and to stick with cooked food and bottled drinks as it was easy to pick up a stomach bug. Atul was a Jain, an old religion that bore similarities to Hinduism and Buddhism. Jains are strict vegetarians which meant that we ate very healthily, though it took some getting used to eating with my hands. I really enjoyed the food. My wife didn't like curry so that was a great opportunity to savour the various spices and different types of curry, although I never got used to the large number of chillies in some of the dishes.

The only sour note during the trip was when we visited a factory in a run-down suburb of Bombay. Atul had placed a small order with them for one of the divisions of the BSC which had placed a repeat order that was currently in production. We were met by the factory owner, a short thickset man with a patch over one eye which gave him a sinister look. I apologised and said that unfortunately we didn't have much time, as we had to get to the airport, and could we look round the factory. We were shown into a dark room where about 20 men were crossed-legged on the floor lasting the sandals. I noticed that one of the workers was putting the straps in the wrong slot. I pointed this out to the owner who picked up the wooden last, inspected the faulty workmanship and proceeded to bang the worker over the head with the last. The worker slumped to the floor, dazed, before recovering and continuing with lasting the sandals. I was flabbergasted. The factory owner looked at me with his good eye and continued to shake his head. I left the factory in a state of shock. Needless to say we didn't deal with this factory again. It concentrated my mind on making sure that certain standards were followed in the factories we dealt with, and that included not hitting employees on the head with the last. I was reassured to hear from Atul that this was exceptional behaviour and he would not tolerate it from our factories. I hadn't seen any evidence of employees being treated badly at the other factories we had visited although standards of employee welfare were very different from our factory in Dalston.

My next visit, a couple of months later to inspect the production, didn't start auspiciously. I had carried with me a large case filled with computer cards from the BSC to give to the factories. This was before the days of barcodes and computer cards were placed in the shoe box for stock control purposes. I was stopped by customs at the airport who wanted to know what was in the heavy case. I then encountered Indian bureaucracy. The officials hadn't seen computer cards before and despite my best efforts to explain what they were there was resistance to me importing these suspicious looking documents into the country. A large gathering of customs officials then ensued to discuss this unusual consignment. Five hours later I was still waiting for a decision to be made. I needed these cards. As often happened, they had been given to me by the BSC at the last moment and unless I gave them to the factories, we would have to open all the boxes again in the UK to insert the cards, a time-consuming and expensive job. In desperation

I offered to pay a "special" duty to resolve the problem. After some negotiation this did the trick, and I finally left the airport. A lesson for the future in how to speed things up.

The factory visits to inspect the production went smoothly. The sandals looked good and as the confirmation sample I had agreed with the BSC. There were two issues that concerned me. The first was a health and safety issue. There was a grinding room in the factories where the wooden wedges were shaped. The wood arrived in blocks and the workers in the grinding room used the machines to chip away the wood to get to the required shape of the wedge. Given that the wedges were shaped not moulded they were surprisingly consistent. The problem with the process was there were wood chips flying through the air and in the faces of the workers. I couldn't stay in the room for more than a couple of minutes. There were no extractors and many of the young operatives refused to wear face masks. The result was a high degree of lung problems among the workers. I discussed this with the factory owners, but they couldn't be persuaded to make the changes. Health and safety issues are taken much more seriously today.

The second problem concerned the colour of the sandals. After the uppers were sprayed with the dye they were placed on the roof of the factory to dry in the sun. Depending on the strength of the sun the colour of the sandal was a lighter or darker shade. The pairs of the sandals matched but different pairs in the order had slightly different colours. Would this be a problem? There wasn't much I could do about the pairs that had been made. In fact, given the process, there wasn't much I could do about the sandals in production either except to make the factories aware of the issue and avoid too great a variation. The hope was that the BSC recognised that this was a natural, artisanal product and that the variations in colour were to be expected and were not unattractive. The issue hadn't been raised with the first order, which had sold well, so I was hopeful that the next order would be accepted without comment. I was beginning to understand that Indian sandals, given their natural characteristics and hand-crafted manufacturing processes, were not an easy product for a developed market like the UK that liked every sandal to look identical.

A few weeks later the sandals were shipped and arrived at BSC's giant warehouse in Leicester. I got a call from one of the quality inspectors saying that there were colour variations. I explained that the sandals were made in a vegetable tanned leather and colour variations were to be expected. Reluctantly they agreed to inspect the delivery and reject those where the colour variation was too great. I complained to the buyer who was sympathetic and agreed to instruct the inspectors only to reject those sandals which were materially different from the confirmation sample. Fortunately, the summer of 1984 was a hot one and the sandals sold well. The buyers were pleased, the factories were pleased, it looked like we were onto a winning formula. The new styles sold particularly well, which was a lesson in being innovative and developing new products so that there is a point of difference. Things then started to go wrong with the Indian sandals. The main issue was that the sandals were failing in wear, and we were getting a high level of returns. The straps on the sandals were unlined leather and after a certain amount of wear were breaking. We also had an issue with the straps coming away from the sole. These were handmade artisan sandals that were not designed for the British weather, or the punishing wear being given to them by the UK customer. The returns were at such a high rate that our modest profit was fast being eroded. It was depressing visiting the BSC and wading through piles of worn, smelly sandals that had been returned. In many cases I didn't feel the return was justified as the customer had had good wear out of the sandal. Unfortunately, this wasn't an argument that the BSC quality team accepted.

For spring/summer 1985 I decided to reduce the styles that I offered to those that were more sturdy and less likely to fail in wear. I considered abandoning the programme completely but decided to give it one last try. I received the orders from the BSC and placed them with the factories. The orders were a lot smaller, but this was sensible given the high return rate. On 31st October 1984 Indira Gandhi, the Prime Minister of India, was assassinated. The story was all over the news. Little did I think that this tragic event would be the final straw in my Indian adventure but bizarrely it was. It was thought that Mrs Gandhi had been assassinated by a Sikh. The wood for the wedges came from the north of India and was transported to the factories by lorry. The lorry drivers were predominantly Sikhs. Because they were being accused of her murder they were being attacked. The drivers went on strike in protest and stopped delivering wood for the wedges. Production was halted for weeks, resulting in some of the orders being cancelled because they would be late. Of all the reasons for a late delivery (production problems, lack of components, strikes, floods, fires etc.) the assassination of a prime minister has been the most unlikely excuse I have given. This delay was a blessing in disguise as it hastened my decision to pull out of India. The return rate continued to climb. We were losing money. I decided to cut our losses.

My first attempt at importing had been a failure. I had learnt a valuable lesson in making sure that the product stood up to the heavy wear of the UK customer. As the importer you were in the middle of the transaction, between the factories and the retailers. The factories blamed the quality failure on the British consumer, who didn't look after their shoes. Typically, they had one or two pairs of shoes that they wore until they fell apart. On the other hand, you had the retailer, who didn't want to upset the customer, and accepted the shoes they brought back without questioning whether they had been subjected to excessive wear. The reality was more nuanced. There were certainly customers who took advantage and returned shoes that should not have been accepted. But there were factory owners (especially Italian ones) who refused to believe that there was a fault with their shoes even though it was patently obvious that there was a mistake. As far as India was concerned the product was too vulnerable and the risks too high. I didn't return to India for more than 25 years.

7. Next stop Italy and Brazil

Around this time, I was introduced to an American from New York called Gary Plotnick. He was close friends with the CEO of a large US footwear brand whom I had met at a dinner party. Gary was a quiet and thoughtful man but also a bon viveur, who loved Europe and enjoyed entertaining his customers at the best restaurants. I greatly enjoyed his company and learnt a lot from him. He had broad interests in politics and culture. But it was our shared love of food that was our main common interest. He was prepared to drive for hours to a restaurant that he read had exceptional food. We were rarely disappointed. In his understated way he was excellent at selling and building strong relationships with the buyers and their bosses. He also taught me what not to do. He overextended himself by opening offices in Spain and Brazil with inadequate funding and management which in the months to come were to cause me serious problems.

His love of Europe and criticism of America caused an embarrassing incident. He had joined my wife and me for dinner at Scott's, the famous fish restaurant in London. He was discussing his view that Americans were naturally conservative, fort>22rt>222

22222

lacking in culture and inward-looking. Sitting next to us was an elderly American couple. In a broad Southern accent the man turned to Gary and told him that he was an insult to America. How dare he speak in these terms about the greatest country in the world. He should be ashamed of himself. The episode certainly put a damper on our meal. Tragically Gary died a few years later of a heart attack while on his exercise bike. He was in his early fifties.

Gary had a footwear agency called DICA that was based in Florence. Italy was still an important producer of both casual and dressy sandals. The casual sandals were made in the area around the ancient Tuscan town of Lucca. At that time there was a strong trend in the US for clogs with a brand called Candies and a flat strappy sandal by a company called Bass that had a padded suede sock. All the factories were making versions of these styles. The dressy sandals were made south of Florence near a town called Arezzo. There was a large group of agents based in and around Florence who were regularly visited by American retailers and importers. Many stayed at the Lungarno hotel overlooking the river Arno, right near the famous Ponte Vecchio in Florence. On my first visit to see Gary I was met at Pisa airport by the man who ran the Florence office, a tall, good-looking and charming Italian called Alessandro Badalassi. He drove us to a picturesque restaurant in a small medieval hill town called Montecarlo near Lucca. There were about a dozen people, all American, seated round this large table in a beautiful garden overlooking the verdant hills of Tuscany. I was greeted warmly by Gary and introduced to the Americans who were customers of DICA. I proceeded to eat a truly magnificent lunch of antipasti, pasta and grilled fish in this magical setting, with a convivial group of Americans. This was the life. A far cry from Dalston or Kanpur. I could certainly warm to importing shoes from Tuscany if this was an example of what it was going to be like.

As it happened, we had limited success importing from Italy. The main problem was the competition, which was intense. A lot of the factories dealt directly with UK retailers, cutting out the middleman like us. We did sell some attractive sandals from smaller factories that Alessandro knew. Driving round the hills of Florence, visiting these small family-owned factories in out-of-the-way villages, were some of the most enjoyable times in the footwear trade. Alessandro often bought a picnic of a fresh loaf of ciabatta, a selection of salami, prosciutto and local cheeses which we ate in the shade of a cypress tree overlooking the hills below. The meal was accompanied by a bottle of a local Chianti. I was warming to the Italian way of life. There was one factory called Framor (short for Fratelli Mora, the Mora brothers) in a village near Lucca where you had to manoeuvre the car through this impossibly tight medieval arch to get to the factory. When I first met Mr Mora and shook his hand, I noticed that two fingers were missing, the result of an accident in the factory. Framor made fine, delicate sandals that were small works of art. We wouldn't be able to make sandals like this in our factory in Dalston. We didn't have the shoe-making skills that had been built up over generations and gave the shoes a unique quality. There was also an infrastructure of suppliers on their doorstep that produced the finest leathers and components that made their job so much easier.

The sandals from Framor sold very well but the capacity of the factory was small so the potential for making much money was limited. The harsh reality is that most of the factories like Framor disappeared from the late 1980s onwards. Apart from those factories making shoes for the luxury market, most of the production of medium- and low-priced footwear moved to the Far East. Some of the more commercial factories moved their production to eastern Europe or north Africa and still managed to put the coveted "Made in Italy" stamp in their shoes by inserting the insock and boxing the shoes in Italy. The tragedy, in this search for cheaper shoes, has been that artisan factories making beautiful shoes have disappeared. Although shoes from China and other Far Eastern countries have improved, they still lack that elusive quality that generations of making beautiful shoes had in their DNA.

I travelled round different areas of Italy visiting factories and sourcing shoes (and enjoying the food). From the Marche region on the Adriatic coast, to the area around Verona in the north, the Naples region and a town called Barletta in Apulia that made cheap sandals. Whenever I told an Italian from the north that I was going to Barletta they always referred to that part of Italy as Africa, illustrating the big divide between the rich north and poor south of Italy. Since then, with the popularity of areas like Puglia as holiday destinations, the divide has become less distinct. We didn't make much money from our Italian imports, which was an improvement on our Indian experience, but it did give me a valuable education in shoe-making techniques, particularly the importance of the finer details. The Italians had a strong work ethic, often staying in the factory until 9 and 10 o'clock at night. They had a passion for making beautiful shoes, a passion that I was starting to acquire. When I founded Dune in 1992, I dealt with a few of the factories I had visited in my travels in the 1980s that had survived the competition from the Far East.

The other attraction to doing business in Italy was Florence, the home of the famous footwear and accessories brands Gucci and Ferragamo. Not only was it a magnificent city with a wealth of Renaissance history and art, and excellent shopping, but it also had some outstanding restaurants offering local Florentine dishes like Bistecca alla Fiorentina, a T-bone steak served rare, that is delicious. It was accompanied by white beans (*cannellini bianchi*) that were bathed in peppery olive oil. We used to leave Florence at the crack of dawn to drive to the factories. In the evening, we would return and have a fantastic meal. The only time I had a disagreement with Alessandro was when I criticised his driving. We were driving back to Florence on a dark and winding road. Alessandro was in a hurry to get back to meet a customer. He was driving very fast into sharp bends. I asked him to slow down. I should have known better than to make a comment about an Italian's driving. It is taken as a huge insult. He was not his usual ebullient self for a couple of days until I apologised, which partially cleared the air.

Following the success of his Italian operation Gary expanded his business and set up offices in Alicante in Spain and Novo Hamburgo, in the south of Brazil. Each was the main shoe-producing area in its respective country. As I have said, it wasn't a wise decision. He ended up overextending himself. The team he had running these offices were not of the same calibre as Alessandro and his team in Florence. Brazil produced good quality leather shoes. The leather came mainly from Argentina where the size and quality of the hides were excellent. About this time, I had taken on a designer called Stanley Dove. Stanley was an outstandingly good designer. He was about Len's age, 15 years older than me. He used to be a manufacturer, but his love was designing so he had closed his factory to focus on being a freelance designer. He was a short stocky man with wiry grey hair, blue eyes and a warm, welcoming smile. He drove a blue Saab 900 convertible with the roof down all round the year. His talent was not only creating interesting and commercial styles but also getting the design and pattern to fit the foot so that it looked spot on. He was keen on music, particularly the old American classics. There was a song written in 1927 by composer Jerome Kern and lyricist Otto Harback called "Smoke gets in your eyes" that he particularly liked. He also liked his metaphors. When designing he was insistent that you had to get the pattern exactly right. Almost right wasn't good enough. In the song the words had to be "smoke gets in your eyes". It would be nearly right but completely wrong for the words to be "smoke gets up your nose". Over the years I have seen too many "smoke gets up your nose" designs.

I visited Brazil to see Gary's office and to better understand the capabilities of the factories. I had flown to Rio and spent an enjoyable day sightseeing and then got an internal flight to Porto Alegre, the capital of Rio Grande do Sul, the southernmost state in Brazil, not far from the border with Uruguay. From there it was a 40-minute drive to Novo Hamburgo. The flight was one of the worst I had been on. Midway through the flight we hit an air pocket. The plane dropped from the sky. The lunch trays shot in the air and food went everywhere. It only lasted a few minutes. The passengers remained surprisingly unperturbed as if it was a regular occurrence.

Novo Hamburgo and the surrounding towns are dedicated to footwear production and associated industries such as tanneries. The countryside was vast, with rolling green forests and large areas growing crops such as soybean, rice and tobacco. The town itself was unremarkable with 250,000 inhabitants. It got its name from Hamburg, the port from which the German immigrants left their country towards the end of the 19th century and settled in the town. Its population is still largely of German descent. The hotel, the Suarez, where I was staying, was basic but clean, much like a Travelodge. I was meeting the team from the BSC who were staying in Porto Alegre at a 5-star hotel. Gary's office was impressive, his team less so. There was a sample making unit attached which was like a mini factory. One of the advantages of the Brazilian offices was that they could make a sample in 24 hours, so the designs you gave them to develop would be ready for inspection the following day. Stanley particularly liked this aspect of Brazil.

The factories I visited were large, usually with two or three long production lines each making 2,000 to 3,000 pairs a day. The quality of the shoes coming off the production line was excellent. There were several operations, such as an automatic buffing machine that gave the leather an attractive rich patina, and a pounder that gave the feather edge of the shoe (where the upper meets the sole) a well-defined border, which we didn't have in Dalston. Most of the production was going to US brands and retailers.

We prepared the office and showrooms for the visit of the BSC buyers. Stanley laid out all the samples we had prepared, and they looked good. As this was their first visit we wanted to make a good impression. We wanted to show that we were working with a professional outfit who knew the market and had a strong relationship with the factories. As it later proved, the relationship with the factories didn't work out too well.

I drove to their hotel in Porto Alegre to collect the buyers. I was told by the front desk to go up to the room of one of the senior buyers. I knocked on the door. After a short delay a scantily dressed attractive young woman answered the door and said, in broken English, that he would be down in about half an hour. After an hour he appeared to join the other buyers, and we drove back to the office in Novo Hamburgo. The BSC team stayed for about half an hour, a quick visit. They seemed satisfied with the operation, although they remained non-committal. The positive was that they liked the samples. Their advice on the factories was that they could be categorised into three groups, A, B and C grade. The A factories were top grade, and the C factories were to be avoided. They then left for what was most likely a more important appointment. Stanley was amazed at their arrogance and lack of engagement. I was more sanguine. I was used to dealing with the BSC. The fact that they had visited at all was positive.

At the start of 1985 Stanley had designed a flat, pointed toe slingback that the Dolcis division of the BSC had bought as a trial from Brazil. We had a phone call from the buyer, a larger-than-life character called Bill Fouracres, to come up to Leicester. I thought to myself: not another quality issue. Unusually, we were told to go straight up to the meeting room. The quality issue must be serious, I thought. Bill came into the room. He was a large, confident man, with a world-weary expression. He was friendly, supportive and had a dry sense of humour. He had an aura of unpredictability. He certainly liked taking risks.

Buyers came with different approaches. Some were indecisive. They were a nightmare to deal with as you never knew whether the order would be cancelled or the style changed, or indeed if

73

you would receive an order at all. Others got too involved in the detail and wanted their shoes to be perfect. Some were passive aggressive, and you didn't know where you stood with them; others were outright unpleasant, and you tried to avoid them. Some liked to chat and get your view on the market and fashion. This exchange of ideas was important as it gave a valuable insight into the market. Others were taciturn and the meeting was over in a couple of minutes. There was one buyer at a retailer called Olivers where if he was in the room for more than a couple of minutes you felt, for no good reason, that the meeting had been a success. Bill was confident and decisive. One autumn/winter season he decided to buy only black footwear. A brave decision that made the shops look drab but commercially astute as black was all the customers wanted. He was in his early thirties but looked a lot older. This was largely because he had a drink problem. His eyes were usually bloodshot. On two occasions when I had previously visited the BSC I heard that Bill had been in a serious car accident. In one of these he had crashed his car into a tree, and it was a close call as to whether he would survive. Fortunately, he did.

He came into the room smiling. The pointed toe slingback had sold outstandingly well. Most of the stock had sold out in the first week it was in the shops. He proceeded to give us a large repeat order and added further colours. He mentioned that because it was so strong some of the other divisions wanted to place an order on the style. Buyers from the other divisions then filed in and placed their orders. When they had all left, we added up the orders. It came to 150,000 pairs, over £2m, a staggeringly large order. Bill came back into the room and gave us a small lecture on how important the orders were and not to let him down.

Excited by the size and potential profit of the order we went back to the factory and sent the order off to Gary's office in Novo Hamburgo. After several reminders we eventually got an email back saying that they were having difficulty placing the order.

It was too large, and the factories didn't have the capacity as they were full of orders from the US. I got on the phone to Gary and pleaded with him to find the factory capacity. This was an amazing opportunity, and we couldn't lose it at this early stage in our relationship with the BSC. He was sympathetic but his team didn't seem to have the clout with the factories to get the orders placed in a suitable factory.

There was a further problem. Around this time the pound had fallen to an all-time low against the dollar. It was down to $1.06 to £1 sterling, almost at parity. This was based largely on the strength of the dollar. When I had taken the order, the pound was much higher. Unfortunately, I hadn't covered my dollar exposure by buying dollars forward. The result was that my already modest margin had been practically wiped out. Despite this, for the sake of my credibility, I had to place the order. I didn't want to tell Bill I couldn't supply the shoes. Gary's office decided that it was worth travelling to a city near São Paulo called Franca, which was another large footwear producing area, although specialising in men's shoes, to see if we could place the order there.

I flew to São Paolo with our quality inspector, a man called Peter, to visit Franca and, hopefully, place the order. Franca was about an eight-hour drive from São Paolo so Gary's people had booked us on a local flight. Peter, who was a nervous flyer, looked at the plane and said he wasn't getting on it. This wasn't a good start. A car was arranged to take him on the long journey to Franca. When we eventually got there, we visited several factories that were all capable of making the order. Their quality standards looked okay, but did they have the capacity? They said they would review the orders and respond the following week with their answer. We flew back to the UK with little confidence that they would accept them. A week later I got a phone call from Brazil confirming that they couldn't find a factory to place the orders. Reluctantly they had to be cancelled.

My next visit to the BSC was not a pleasant one. I was kept waiting for a long time. I explained to Bill that despite my best efforts I couldn't place the orders. I felt like a complete idiot. I had been given a great opportunity and had spectacularly failed to deliver. To be fair he could have been more difficult. Despite his threats to charge me with loss of profit he didn't. Instead, he gave the orders to a man called Verno Kirsch, the owner of one of their main Brazilian suppliers called GVD, who placed the orders with little effort. The shoes were a great success, but we didn't reap any of the rewards.

After the Brazilian fiasco I decided to stop our importing business to concentrate on manufacturing. We closed Capital Shoes. Despite all the effort we had ended up breaking even on the venture. It could have been a lot worse. I had learnt a lot from the experience which was to put me in good stead for my next business. What had I learnt? Make sure the quality of the product is good and will stand up to a reasonable amount of wear and tear. Build in enough margin for returns as some of the shoes will fail. Deal with factory owners and agents who know the market, have strong relationships with factories, and are trustworthy and honest. Develop a range that has a point of difference so that you are not just competing on price as there is always someone who will beat you on price. Cover your currency risk by buying the foreign currency forward so that your profit is fixed. If there are problems, and there always are, you need to be honest with your customers. They are a lot less sympathetic if they know about a problem (late delivery, quality issues, the assassination of a prime minister etc.) when it is too late to do anything about it. Trust is an essential element of the relationship – indeed any relationship.

Meanwhile the situation with the factory was getting no better. We were making a profit, but it was shrinking year by year as more and more production went overseas. Len was still committed to the factory and after my foray into importing, I was committed to supporting him as much as I could.

8. First trip to Far East

Although my experience of importing had not been a great success, I was becoming more and more aware that manufacturing fashion footwear in the UK had no future. London Lane Shoes was struggling to make a profit; margins were being squeezed and orders getting smaller. I was approaching 40, we had two young children, and I needed a business that had growth potential where I could earn a reasonable living, and that wasn't making women's fashion shoes in Dalston. There was the option of leaving footwear and focusing my energies on another business, but nothing obvious came to mind. I had also invested the past 10 years in the footwear industry. I had a good understanding of the challenges, understood how shoes were made, knew a lot of the people in the industry, really enjoyed the product and, as everyone kept telling me, people will always need shoes (although in some of the lean years that followed, I did question this assumption). Going into retail was an option but that would require a lot of capital. Seeing a raft of specialist footwear retailers struggling didn't encourage me to pursue that avenue. What finally persuaded me to leave footwear manufacturing was a trip to Taiwan.

Every June and January a footwear fair takes place on the shore of Lake Garda at a small town called Riva del Garda, a popular holiday resort and a picturesque location in north Italy near the Austrian border. I regularly visited the fair looking for ideas and to get an understanding of what was happening in the market. While I was wandering around the fair in June 1985, I came across two Far Eastern companies offering what looked like good quality court shoes. One was called Sunny Shoes. It was owned by the Yang family and had factories in Taiwan. One of the brothers, Jerry, who spoke good English, was managing the stand at Garda. Although the lasts and shapes were not quite right for the UK market the quality and the prices were attractive. The other was a Japanese company, Sinco, which had an office in Taiwan. Its range looked better quality than most of the Far Eastern stands. Although a lot of the shoes were in fabric or synthetic materials, they had an impressive attention to detail. Both Sunny and Sinco were keen that I visit them to see their operations. I took their business cards and said I would contact them when I got back to London.

In the 1980s the two main producers of footwear in the Far East were South Korea, which made predominantly sport shoes and trainers, and Taiwan which made more dressy formal shoes. Hong Kong was the third main source, but it primarily made cheaper canvas and rubber products. Japan had been a major producer during the 1960s but most of its production had moved to Taiwan and South Korea where the large Japanese trading companies had encouraged local entrepreneurs to start footwear production. It wasn't until the late 1980s that production started to move to China when the Chinese government opened 14 coastal cities as manufacturing centres and allowed foreign investment. The southern province of Guangdong, just north of Hong Kong, was particularly attractive given its proximity to Hong Kong. The combination of easy access, low costs, and particularly a large, disciplined workforce with good hand/eye coordination, made China an attractive destination for Taiwan,

South Korea and Hong Kong manufacturers to relocate their production. The fact that the typical hourly pay in China at the time was US$0.59 whereas it was US$7.80 in Italy made the move to China particularly compelling.

I got back to London and decided to make a trip in October 1985 to Taiwan to meet with Sunny and Sinco and a few other companies that I had been introduced to. There were no direct flights to Hong Kong in those days, so I flew to Bahrain and then on to Hong Kong. The main airport was Kai Tak, a technically demanding airport for pilots. It couldn't be approached on auto pilot as there was a sharp right hand turn to the runway that extended into Victoria Harbour. The approach was also near the top of buildings which added to the difficulty. The airport had a history of plane crashes, because of bad weather or the demanding approach. I remember on one visit seeing a China Airlines plane floating in the harbour having overshot the runway in bad weather, which wasn't very reassuring. We managed to land safely. I got a taxi to the Regent Hotel in Kowloon where I was staying. I had planned to stay in Hong Kong for a couple of days where I had appointments with some US footwear agents and then fly on to Taipei, the capital of Taiwan.

Hong Kong was a hive of activity. Lots of cars, people and buildings all crammed into a very tight space. The red double-decker London buses were a reminder that this was a British colony. The Regent Hotel was magnificent with a memorable view over Victoria Harbour from the glass-fronted lounge. You could see Hong Kong Island on the other side of the harbour with its imposing skyscrapers leading up to the Peak at the top of the city. Large container vessels and boats of every shape and size cruised up and down the busy waterfront. I was warmly greeted at the check-in desk with the words "Welcome back Mr Rubin. Your suite is ready for you." I wasn't sure they had the right Mr Rubin. They may have confused me with another Mr Rubin. Most likely Stephen Rubin the owner of Pentland and Reebok, the sports footwear brand. Before I had time to say anything,

I was swept up to my suite, a huge room with great views over the harbour. An ice bucket with a bottle of Moët & Chandon champagne and a bowl of exotic fruit were on the desk.

The next day I visited shoe factories. The commercial/industrial areas were only about half an hour from the centre. The factories were typically small units on an upper floor of a busy multipurpose building housing all types of manufacturers and trading companies, from dresses to handbags, from textiles to footwear. They all seemed to have old-fashioned and rickety goods lifts. One of the factories was making fine dressy sandals and shoes with snakeskin uppers. Called King Kong Leather Factory, it was run by a Mr and Mrs Kong. Many years later the King Kong factories in China and Bangladesh became major suppliers to Dune and have been for the past 30 years. I was struck by the entrepreneurial drive of these factory owners. There was a tremendous energy and enthusiasm to do business which was infectious, and which I hadn't found in the other countries I had visited, certainly not in the UK. This was the first of many trips to Hong Kong. Most of them were productive and enjoyable (until I got arrested, but that story is for a later chapter.)

The flight to Taipei, the capital of Taiwan, was only an hour and a half but the difference in atmosphere between Hong Kong and Taiwan in 1985 was immense. The first sign I saw as I entered the customs hall was "Death to drug traffickers". The customs officials were unfriendly, bordering on aggressive. One had the sense that this was a country under martial law, which didn't particularly welcome visitors. The country was run by the KMT (Kuomintang), a nationalist party set up by a strong leader called Chiang Kai-shek who had fled China during the civil war with many of his followers in 1949 and set up his oppressive government in Taiwan. During what was called the White Terror some 140,000 people were imprisoned or executed for being perceived as being pro-Communist and anti-KMT. What the KMT had achieved was rapid industrial and technological growth. Taiwan, Hong Kong, Singapore and

South Korea were known as the Four Asian Tigers for their dynamic economic expansion.

I got a taxi to my hotel. Driving through Taipei I was struck by the large number of gaudy neon flashing adverts. The city felt different to Hong Kong. The influences were more from Japan and the US. There was an energy and lots of activity fuelled by the successful growth of the economy but there was also a sense of impermanence because it was a country that was still finding its identity. The Howard Plaza, where I was staying, was a comfortable modern hotel in the centre of town. It had the worst soundproofing of any hotel I have stayed in. You could hear everything that was happening in the neighbouring rooms which wasn't always pleasant listening. I had picked up a cough during my travels which I had difficulty suppressing and was aggravated by talking. Despite taking cough medicine, it got progressively worse to the extent that I couldn't speak without having a coughing fit. As I was planning to be in Taiwan for three weeks, and didn't fancy spending all this time coughing, I went to a hospital recommended by the hotel. They gave me a course of antibiotics which they insisted I took with some medicine to protect my stomach. The treatment was professional and fast and after a few days, to my great relief, the cough started to subside.

In the first week, in between coughing, I visited several footwear agent's offices, mostly run by Americans. The product ranges were generally uninspiring, with lots of cheap synthetic sandals. With the combination of a debilitating cough, lack of a good night's sleep, being away from home in a city that was very foreign and not over-friendly, and the disappointing footwear I began to question whether the trip was worthwhile.

Matters improved when I visited Sunny Shoes which had offices and showrooms in the centre of Taipei. I met its CEO, Jason Yang. He was the brother of Jerry whom I had met in Garda. He was a successful and ambitious businessman. Various awards, including for Taiwan Young Businessman of the Year, were on display in reception. Jason was a slight man with the look and demeanour of a banker or politician, formally dressed, with a serious yet welcoming expression. He had a quiet confidence and authority. He showed me around their impressive showrooms. He explained that Sunny worked with several factories in Taiwan where they had a strong relationship and were confident in the quality. Their customers were mainly in Europe, the biggest being a large mail order company in Germany called Reno. He was keen to build his business in the UK, so my visit was timely. The shoes in the showrooms were well made but not right for the UK market. They lacked a certain elegance and refinement that characterised the type of shoes that we sold from the factory in London. However, the quality and workmanship were good, so it was a matter of developing new constructions and patterns for the UK market. One of the areas where London Lane was particularly strong was court shoes (or pumps as they are called in the US). I had brought with me an unlined court shoe that had sold well and asked Jason to develop a sample. It was a mid-heeled stiletto heel with a short, pointed toe with slight walling around the toe called Hazel. The cut of the shoe was important. It was what we called a "sexy" cut that showed the top of the toes (the toe cleavage!). The Germans preferred a higher cut profile that fully covered all the toes. He agreed he would make the sample quickly. I arranged to meet him in a week's time at one of his factories in the south of Taiwan near a town called Tainan.

From what I saw in the offices, and later the factories, Taiwan and Hong Kong had a strong work ethic, stronger than in the UK. Most offices and factories were open six days a week. The office workers were at desks that were a lot smaller than in the UK and were partitioned off to give privacy. At lunch time the lights were switched off and the workers put their heads in their arms and fell asleep for a half hour's cat nap. There wasn't much chatting between workers. It seemed to be discouraged. Jason introduced me to other members of his family, many of whom had key management positions in the company. His cousin, Linda, was an efficient and serious lady, who became my main contact. His brother, Johnstone, a tall lanky man, didn't seem to have a specific role but was there to look after customers. Johnstone would drive me to Taichung, the main footwear producing area, halfway down the west of the island and the second largest in Taiwan, where I had several appointments.

The next day I was met by Johnstone and we set off on the two-hour drive to Taichung. He was a fast driver and spent the journey weaving in and out of lanes of the crowded motorway which appeared to be an accepted way of driving. We arrived at the large National Hotel in the centre of Taichung which was to be my base for the next two weeks. After checking in and unpacking Johnstone said that the spa at the National Hotel was famous. He felt it would be good for my cough to have a massage and treatments, which he said had many therapeutic qualities. As I had no appointments that day I accepted. We arrived at the spa where my clothes were taken from me and hung in a locker. After stripping off I was shown into a wet room where two large men threw a bucket of hot and then cold water at me. I was then told to lie on a bed and had a special salt solution rubbed into my body which was then scrubbed off with rough brush. At the same time someone cut my toenails and fingernails. I had more water thrown over me. I was then shown into another room where a large, muscular lady proceeded to pummel my back and give my body a very painful massage. I resisted the urge to shout out from the pain she was inflicting on me. I was then shown into a large bathroom-like area where I was hosed down again, had my hair and body washed, rinsed and finally placed in a large bath robe, escorted to the "recovery" area where I was placed in a large relaxation chair, given a mug of green tea and had my feet massaged. After a short nap I went to get dressed. An attendant was waiting to help me. He handed me my clothes from the locker. My shoes had been cleaned and trousers pressed. I felt surprisingly good after this ordeal. Johnstone insisted we had a bowl of beef noodle soup in the coffee shop, also a speciality of the hotel, that also had therapeutic qualities. Like a lot of Chinese, he slurped the soup very noisily, clearly enjoying its hearty mixture of a strong broth, soft chunks of beef, long noodles and chillies all washed down with a bottle of Taiwan beer. It was very tasty and nourishing. I slept well that night.

The pattern of the second week was much like the first. I visited lots of offices and a few factories but there was nothing of great interest until I arrived at Sinco's offices in a house with a goldfish pond in the front garden. Goldfish are a good omen in Chinese culture because in Asian mythology they are related to dragons, and dragons represent good luck and power. As it happened, meeting the lady who ran the Sinco office, Janna Chen, was one of the best pieces of good fortune I had in business.

Janna ran the Sinco office in Taichung with her colleague, Mike Wu. She had a warm smile that made me feel welcome. I got the impression that she was serious about wanting to do business with me. She explained that her relationship with Sinco was strained. Her small team in Taichung operated independently, although she did have a couple of Japanese customers and visited Tokyo regularly. I was impressed by her footwear knowledge and her enthusiasm. The range in her showroom had one or two interesting shoes but, like Sunny, it wasn't right for the UK market. We visited some of the factories she worked with. She clearly had an excellent relationship with the factory owners and there seemed to be a mutual trust and respect. The factories themselves conformed to a standard format. One long production line (or several for the larger factories) where the uppers and components came together. The manufacturing process was like London Lane, except that as it was on a long conveyor belt, there were more operatives, more shoes being made and the production flowed more easily. There were also long drying tunnels to heat-set the shoes so that they kept their shape. The workers were mainly young women (not underage) who worked fast and efficiently. There were several quality inspections during the production process to ensure the shoes were consistent and met the standard of the confirmation sample. I was struck by how simple the whole process was (much simpler and more organised than our factory in London) and the quality of the final product.

I left Janna confident that she had a well organised operation, good factories and a strong appetite to do business. The only issue was that most of the shoes she was making were fabric or synthetic and I was looking for leather shoes. I left her some samples to make in leather to see how they were executed. I learnt on my return to London that she had separated from Sinco and had set up her own company with Mike called Maxgreat.

Although my cough had improved, I was missing home and was looking forward to returning to London. I felt guilty being away from the factory for so long, although being away had given me a good opportunity to reflect on the challenges we faced maintaining production in London, especially after seeing how the shoes were being manufactured in Taiwan. Not only was the quality of a high standard but the costs were considerably lower. My last appointment was to meet Jason at the factory near Tainan. It was a small unit in a new building and everything was spotless and well organised: tidier than our factory in Dalston. Having a large order for one style certainly helped. Like the other factories the workforce was mainly young women who were smartly dressed in blue workwear bearing the factory's logo. An unlined leather court shoe was on the track. I was struck by the consistently high quality of the shoes being produced. They looked as good as, if not better than, the ones we were making in Dalston. But the production price was less than half what it cost us in London. Jason gave me the sample he had made of Hazel. It was better than the original. It was a defining moment in my visit. I realised, with certainty, that our days producing shoes in London were numbered.

When I got back to London I sat down with Len and discussed what I had seen in Taiwan. The visit had confirmed my view that the future for London Lane was bleak. We couldn't compete with the footwear coming out of the Far East. There was possibly a niche making short production runs of high fashion shoes that we could deliver quickly to the market, but we needed to have a major rethink of the structure of the factory. Len was sympathetic but still believed there was a future in the business. He wasn't prepared to give it up. We decided that he would buy my shares for a modest sum. He would continue making shoes, because that is what he knew and enjoyed, and I would leave the company to start my own importing business. My days as a manufacturer were over. London Lane Shoes survived for two more years before Len finally closed the company.

9. Start Browning and commit to importing

The decision to leave manufacturing in 1986 had not been difficult. I felt sadness at leaving London Lane Shoes and working with Len. We had very different aspirations, but I had huge respect for his work ethic, loyalty to his team and integrity. I had learnt a lot from him not only about footwear but also how to conduct yourself in business.

Having left manufacturing, it made sense to become an importer/wholesaler. In 1986 I registered the company that I would trade through. I lived in a small street in Maida Vale, West London, called Browning Close; I chose the name Browning Enterprises Limited. I felt, most likely misguidedly, that the inclusion of the word "Enterprises" gave the impression that the company was bigger and more important than it was. Calling it Browning Footwear Limited would be too limiting if I wanted to diversify later.

The next step was finding an office or showroom. I didn't want to work from home. We lived in a small townhouse with two young children. There was no spare room, and I didn't fancy working from the kitchen table. I needed an office where I could work without distractions. When I was driving home one day I stopped at a set of traffic lights on the junction of St John's Wood Road (famous for Lord's Cricket Ground, the home of cricket) and Maida Vale. Facing me was a Lloyds Bank and on the first floor there was a sign saying office space to let. The location, 37A Maida Vale, was ideal as it was round the corner from where I lived. The main office was a secretarial agency, and they were letting a small room at the rear of the office that was surplus to requirements. It was small and dark. On the plus side it was cheap and convenient. I took it and kitted it out with a desk, a chair and some shelves to show off the shoes.

The name and the office were the easy part. The harder part was finding the right products and then paying for them – then getting paid myself. Fortunately, thanks to Capital Shoes, I had some experience of importing, and I had learnt from my mistakes. India had taught me the importance of quality and making sure that the product not only looked good but also stood up to wear (and not to deal with factories that treat their employees badly). Italy had shown that it was difficult to compete in a market where the manufacturer could easily go directly to the retailer. Unless you could develop a product with a point of difference, which wasn't always easy, your margin would be under pressure. Brazil, like Italy, was a well-developed sourcing destination. Companies like GVD (which was the recipient of the order for the flat, pointed toe slingback that I couldn't fulfil) had a strong foothold in the UK market. The Brazilian office I had worked with was not good enough, and as a result we had lost a large order and damaged our credibility. I knew I had to find a good partner, whether a factory or sourcing office. One I could trust, who knew the market well, had strong relationships with the factories, They needed to understand the importance of meeting quality standards, communicated honestly and quickly (it was always better to get bad news sooner rather than later; unfortunately, there was always bad news) and was committed to the partnership. The trip to Taiwan had opened my eyes to the opportunity of sourcing a better grade of product which, as far as I could see, no one else was offering in the UK. Most of the imports from the Far East in the 1980s were cheap basic products. From my trip I was confident that I could develop a range of premium products at competitive prices. If Sunny or Maxgreat (or both) executed my designs correctly, and if they were serious about the partnership, I was confident that there was an exciting business in prospect. The criticism of the fashion products from the Far East was that they were too exact and lacked any intrinsic character or personality. My challenge was to give the shoes that X factor, which was all about the cut of the pattern, the feel of the material, the flexibility of the sole and a certain *je ne sais quoi* that made the shoe special.

My business model would be based on adding value through design, building strong relationships with both the customers and the factories, working to make the supply chain as fast and efficient as possible and never compromising on quality. With those four pillars and, over time, building a strong team, I would (barring disasters outside my control and there were a few of these) create a viable business.

I wasn't a natural salesman but found that, even though I wasn't an extrovert, if I developed certain qualities, I could succeed. What were those qualities? Firstly, I had to be passionate about the product. Over ten years I had become a shoe obsessive. To the annoyance of my wife, I was constantly photographing shoes in shop windows to get ideas. Although I wasn't a designer, I could sketch well enough to communicate my ideas to the buyer or product development team at the factory. I knew when the shoe was right and when it wasn't. Having worked with the likes of Stanley Dove at Capital Shoes, the designer Sue Saunders at London Lane, and some of the talented buyers at the British Shoe Corporation, I understood the importance of attention to detail, of not compromising but making sure the shoe was spot on. A few millimetres could be the difference between a shoe selling well and being average. If it sold well there could be repeat orders and, just as important, my reputation as a supplier would be enhanced.

The second quality was credibility. The customer and the factory had to trust me. They had to be confident that they were dealing with someone who wasn't going to let them down. Someone who would keep them fully informed with both good and bad news. Buyers are much more sympathetic if you tell them the shoes are going to be late well in advance rather than a few days before delivery. Many of the problems I was to encounter in the years ahead were because factories or the buyers had promised something that they couldn't deliver. After a while I could sense that what I was being told just wasn't realistic or credible. The skill was how to manage that situation without undermining my own credibility.

Thirdly, I had to give a fast and efficient service. That meant not only communicating quickly and accurately so I wasn't wasting the buyer's time, but also looking at ways to improve the speed of delivery. The time it takes between the customer placing an order and delivering it into their warehouse varies depending on the country making the shoes. From the Far East it can take five months whereas from Europe it can be five weeks. The big advantage of getting the shoes from the Far East is the much lower price and higher margin for both you and the retailer. Unlike clothing, footwear has long lead times. The large number of components and processes in making a pair of shoes, the limited capacity of the factory and the long shipping and delivery time makes it a lengthy process. If you can reduce the lead time by a few weeks, it can have an important impact on the retailer's profitability. They can turn their stock quicker (which is good for cash flow), get faster customer reactions and have the opportunity of placing reorders. Finally, I had to be persistent, a much-underestimated quality. There is a balance between stalking a buyer and being persistent. If I was to succeed I had to keep communicating with the customers so that they knew I wouldn't give up.

The main challenge was how to fund the business. I had little capital. I had left London Lane with very little. I had taken out a mortgage to buy our house. There was the option of becoming an agent of the factory/sourcing office rather than the principal. This had the advantage that the factory/sourcing office funded the sales and took the risk, and I would receive a commission which I would negotiate, typically around 10 per cent. The disadvantage was that I would have to rely on the factory to pay me. I had heard stories of agents either struggling to get paid or more frequently having their commission reduced because there had been a quality problem, a discount had been taken by the customer for late delivery or some other reason that had not been the agent's fault or responsibility.

Another important consideration was that the margin you could make as the principal could be significantly higher than an agent's commission, especially if you found a source, as I had found in Taiwan, which was not only competitive from a price point of view but also produced a high-quality product. The biggest downside, as I was later to find to my cost, was if there was a quality problem or other claim, which meant that payment was withheld by the customer. In those circumstances you had to rely on the factory to be reimbursed. If you had paid by an irrevocable letter of credit, as I would need to, as my company had no trading history or capital, getting paid by the factory would depend on them taking responsibility for their product and honouring a fair claim. The letter of credit was a document issued by a bank guaranteeing payment if the documents required by the letter of credit, such as an invoice, packing list and bill of lading showing that the goods had been shipped, were presented on time and in the form prescribed. I opened a lot of letters of credit, as this was the standard form of payment in the Far East until, some years later, I achieved a level of profitability and credibility which meant that I could trade on an open account without the need for a letter of credit or other bank guarantee.

For me to start trading I needed a letter of credit facility from a bank. The bank required financial information that showed that the business was going to be profitable, together with a cash flow forecast showing how much money I needed to operate the business. I drew on my accounting knowledge to prepare a five-year business plan that detailed the budgeted profit and loss, balance sheet and cash flow forecast. The plan wasn't very complicated as the business model was simple. It showed, like all good plans, that if all went well, I would make a healthy profit. The way it worked was that I would receive an order from the customer at the sales, or wholesale, price. I would place a matching order with the supplier at the cost price. The letter of credit would be used to pay the supplier when the goods were shipped. I would invoice the customer when they were delivered to their warehouse and I would be paid, usually 30 to 60 days after delivery. It could be six months or more from the time you received the order, opened the letter of credit to the factory, got the order manufactured, shipped and delivered to the customer before you were paid. It was a long process that needed a generous facility.

It was important that the margin (the difference between the wholesale price to the customer, after any discounts, less the cost price paid to the factory/sourcing office plus the costs of importing the order such as freight, duty and delivery) was acceptable. I aimed for a 25 per cent margin on the wholesale price from Taiwan, and the small business I was doing from Spain and Italy would be more like 12 per cent. As the years went by this margin got lower and lower as trade became more competitive until some 20 years later, we were struggling to make 10 per cent from the Far East. I was planning to keep my overheads to a minimum. Initially I was going to do all the administrative and operational tasks myself.

I knew that HSBC specialised in trade with the Far East so I made an appointment to meet it and another Far East specialist bank, Standard Chartered, to see if they would give me a facility.

Standard Chartered turned me down on the grounds that I was a start-up. At HSBC I met a charming woman, called Linda, who was extremely helpful. Given my extensive knowledge of the footwear industry and experience in manufacturing and importing, even though this was a new company, she was prepared to give me a small letter of credit facility. The fact that I was a chartered accountant was an important factor in persuading her that I had good commercial awareness and was a risk worth taking. The letter of credit relied on Browning's customers' creditworthiness as HSBC had a charge on the amount it owed. There was, however, one important condition. As the company had no capital, apart from the share capital of £100, HSBC required a personal guarantee secured on a second charge on our house. I wasn't happy to accept this condition. Risking having our house taken from us was a real concern. Unfortunately, I didn't have a choice if I wanted to start trading, so, reluctantly, I accepted the second mortgage. Linda's advice was to get an investor who could put some capital into the company. If that happened, they would reconsider the requirement for the personal guarantee. My later experience with banks was that once you have given a personal guarantee it is a struggle to get it released.

I needed orders from retailers, so I spent a lot of my time on the road visiting customers. I would make a weekly return trip of 220 miles up the M1 motorway to see the buyers at the British Shoe Corporation in Leicester who I knew would place the largest orders. Every week I would return without an order. There was no enthusiasm to take on new Far Eastern suppliers as the BSC had its office in Hong Kong which was its preferred source. I kept on driving up the M1 week after week, showing them my samples.

Instead of the BSC, my first order came from Olivers, a large multiple footwear specialist based in Leicester. It was not, as I had expected, for the court shoe from Taiwan but a flat, fabric casual from Spain. I had been to Alicante a few times and met an agent called José who had been recommended. He was a short tubby man who smiled a lot and was brimming with enthusiasm. He knew the market well and had good contacts with the factories. The two main shoe-producing towns near Alicante are Elda, whose factories specialise in high grade dressy shoes, and Elche, which is known for cheaper, more commercial footwear. José had introduced me to a small factory in Elche that specialised in vulcanised fabric casuals.

Vulcanised rubber soles were a process of baking the rubber and other materials in an oven so that it bonded to the upper and produced a light and flexible sole. The owner was a young man called Manuel. His factory was small but well organised and the shoes he was making were good quality and commercial. The buyer at Olivers liked the product and placed an order for 10,000 pairs. As this was our first order with Manuel (and for Browning) I went to Spain to inspect the shoes. They looked excellent. The upper was in an attractive cotton knit fabric and the rubber sole suited the style. I had the shoes tested at SATRA (Shoe and Allied Trades Research Association) which was a specialist research and testing company. It gave the shoe a clean bill of health. My first Browning order was delivered in March 1987.

A few weeks later at the Dusseldorf shoe fair I saw the Olivers' buyer striding towards me. He was a tall, serious man. His expression as he approached was particularly stern. There was bad news, he said. The fabric shoes we had delivered were all tearing down the centre of the vamp. There was a fault with the material. He would have to return them all. My heart sank. I had opened a letter of credit to Manuel's company. Would he meet the claim? Did he have the money to meet the claim? His was a small family business that operated hand to mouth. On the plane back to London I reflected on the consequences of this setback. A lot of dark thoughts went through my head. If the letter of credit went through without an issue, and there was no reason it shouldn't, if Manuel couldn't meet the claim, if I couldn't repay the bank and they called in the personal guarantee I could possibly lose my home. What was I doing putting my family home at risk? Even after all the inspections and tests the shoe had failed due to some freak fault in the material that clearly had not been present when it was tested by SATRA.

I got back to London. I questioned whether the fact that my first order had failed was an omen, that importing shoes was too risky and ultimately a huge mistake. I didn't want to scare my wife so I kept this to myself. The next day I phoned HSBC to ask if the letter of credit had been presented. They confirmed that it had been and that they were about to phone me. There were several errors in the documents that had been presented with the letter of credit. Was I prepared to accept these errors or return the letter of credit to the presenting Spanish bank, refusing payment? Miraculously I had a lifeline. I refused the letter of credit. I phoned José to explain the problem. Given the impact this would have on Manuel, he asked me to fly to Spain to explain the problem to him and bring some examples of the failed shoes with me. The meeting with Manuel was one of the most difficult ones I have had with a supplier. Although he accepted there was a fault with the material, he was distraught. This would ruin his business. He had a young family. Couldn't I help him? He started to cry. I tried to explain that there wasn't much I could do. Couldn't he claim against the material supplier? I left the factory feeling terrible. Even though I felt I had behaved entirely properly I still had a nagging feeling of guilt. It had been a close call. In the end I concluded that I had been both very unlucky (the material inexplicably failing) and very lucky (the errant letter of credit). I needed to press on and devote my energies to Taiwan.

The sample of the Hazel unlined leather court shoe had arrived from Sunny and it looked good. I showed it to a couple of retailers, Stylo Barratts in Bradford and Lennards in Leicester. They loved the shoe, and both placed an order. Court shoes were a strong fashion staple and the price at £14.99 was competitive for a leather court. I was understandably nervous. What would go wrong with my second order? I placed the order with Sunny. A couple of weeks later I received the confirmation sample which both Stylo Barratts and Lennards confirmed. The production had to be exactly like the confirmation sample. In July I flew to Taiwan to inspect the shoes. I wasn't taking any chances. I wanted to see the shoes with my own eyes to check that, not only did they look good and were as the confirmation sample, but also that there were no obvious faults with the manufacturing process; that the heels weren't going to come off and had been attached with the right quantity and length of nails, the soles were secure and attached using the correct adhesive and the right pressure had been applied. I had prepared a long quality checklist to satisfy myself, as far as I could, that the shoes would stand up to reasonable wear. After the disaster with the fabric casual from Spain it was essential that these orders were trouble free.

After spending the night at the Regent Hotel, I got the early morning flight to Taipei. I was met at the airport by Johnstone and driven to Sunny's office. I had brought with me some sketches and magazine tears of styles I wanted Sunny to develop. I spent a couple of days with Linda and their designer working through the samples and specifying the constructions and materials. After visiting the factory to inspect the production I would return to review the samples that had been made and make any adjustments. I planned to take the samples back to London to show them to customers. It was a process that I would go through six times a year for the next four years.

I had checked into the Howard Plaza Hotel again as it was near Sunny's office. The soundproofing was still an issue. With loud throat clearing on one side and sexual activity on the other, I didn't get much sleep. I had also been taken by Johnstone to a night market, one of the local tourist attractions. He had persuaded me to drink snake blood as it was meant to improve the immune system and increase virility. I'm pretty sure it didn't improve either of these, but it did give me severe indigestion. In my naivety I was worried about offending my hosts if I didn't try the local delicacies. It took a year or two before I decided to

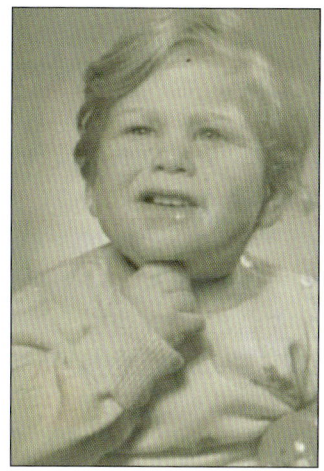

Top left: Me aged two

Above: With my father Louis at our house in Hampstead Garden Suburb

Left: Morris Rubin, my grandfather

Below: With my parents Louis and Dorie at my barmitzvah in 1960

Above: the lasting room of a shoe factory circa 1900

Far left: slipping the upper off the last

Left: buffing the upper to get a shiny finish

Right: a 1950s advertisement for Saxone, part of the British Shoe Corporation, which closed in 1998

Above left: Graduating from the University of Kent in 1969

Above: Me and my uncle Len Goodman (right), whom I went into business with, and a family member, in 1972

Left: Opening our store in Lakeside in 1996 with EastEnder Patsy Palmer

Below: Comme Il Faut – my first foray into retailing

Above: Dune's first store on King's Road, London, which opened in 1993, with shoes on top of boxes

Left: With the Dune team, winning the Multiple Retailer of the Year at Drapers' awards in 2014

Above: The Dune store in Stratford, east London, with the reverse catwalk concept displaying shoes on the ceiling

Right: An advertising campaign in 2015

Above: Three important suppliers: John Khuu (King Kong), Jones Ng (Minglo) and Janna Chen (Maxgreat)

Left: Welcome committee during a visit to a factory in China, 2002

Below: Opening Dune's flagship store in Dubai Mall with Nilesh Ved, CEO of Apparel Group in 2018.

say "no" and stopped eating some of the less attractive food that was offered such as sea slugs, a nasty slimy fishy substance that tasted as unattractive as the name suggested.

The next morning Jason picked me up from the hotel and we drove the four-hour journey to the factory in Tainan to inspect the orders. Any nerves that I had were soon dispelled. The shoes looked excellent. The leather they had used was top quality. Unlike a lot of leather in the market that looked too perfect, almost like plastic, this leather had a nice natural grain and although soft to the touch had kept its shape, which was important for an unlined shoe. Because a plain court is so simple any blemish in the leather, broken topline or poor lasting is obvious. There's nowhere to hide. Fortunately, I couldn't see any faults. I spent the day checking the quality including making sure the labels and packaging were correct as this was an area where factories often slipped up. I checked that the heels had been attached securely and all the other tests on my extensive checklist. Everything passed with top marks. As on my first visit I was struck by the simplicity and efficiency of the manufacturing process. The young women on the production line made it look very easy. I was also impressed with the quality control procedures with a team of operatives checking every pair to ensure standards were met. I left the factory with a sense of relief, hoping that nothing would go wrong between now and when the shoes arrived in the customer's warehouse.

The samples I had detailed were waiting for me when I returned to the office in Taipei. They weren't all perfect but there were a few where they had interpreted my sketches accurately. I hope I wasn't being over-optimistic but they looked like potential winners. As Johnstone drove me back to the airport I felt, for the first time, that Browning was heading in the right direction. The real test would be two months' time when the shoes arrived in the shops. It would be then up to the buying public to give their verdict on Hazel.

The good news came from Stylo Barratts' managing director, Michael Ziff, who called me to say the shoes were flying off the shelves. They wanted to place a sizeable repeat and add three further colours, and could I make a high-heeled version of Hazel. I had a similar phone call from the Lennards' buyer.

Finally, after six months of weekly drives, the British Shoe Corporation gave me an order. In the end the Dolcis buyer took pity on me, either that, or because of the frequency of my visits he thought I was a supplier already, placed an order for a leather court shoe. I think the buyers realised that I really wanted their business, recognised that I had some interesting products and appreciated that I wasn't going to give up. My persistence had paid off.

Things were looking up. I now needed to follow HSBC's advice and find an investor to fund the growth.

10. Funding the business

My success was now threatening to become a problem. The orders I was receiving for style Hazel, and the potential for further orders on the styles I had developed with Sunny, meant that my letter of credit facility with HSBC was inadequate. I needed a bigger facility and to get it I needed more money in my account. I had to find an investor in exchange for shares in the company. I was not happy giving away equity at this early stage as, given Browning's limited trading history, the valuation would be low. But I didn't have a choice. The banks I approached were not prepared to give a long-term loan as I was, in their eyes, a start-up without sufficient creditworthiness. I contacted several private equity companies. Most of them came back quickly to say that they weren't interested because Browning was too small.

One company that did suggest a meeting was 3i, a large private equity company that invested in smaller businesses. 3i had been formed in 1945 by the Bank of England and the major British banks to provide long-term investment funding for small and medium-sized enterprises. Armed with my five-year plan and wearing my best suit I met 3i at its large 1960s building opposite Waterloo Station. The meeting was like a job interview. I sat opposite a team of four executives who asked questions about my career, the footwear industry and my plans for Browning. I left them my business plan and they said they would get back to me. It was difficult to assess whether they were interested, but the meeting seemed to have gone well. A couple of weeks later they came back with a long list of further questions about the business plan. This was all part of their extensive due diligence. I sent back my responses. They asked further questions. Finally, I received an offer document for them to buy 30 per cent of the shares of Browning for £40,000 and to provide a subordinated loan of £100,000 repayable in five years' time. The documentation was long and complicated with lots of restrictions and requirements that gave 3i good protection and me a lot of work.

I next went to see Stephen Rubin, my namesake but no relation. He owned a company called Pentland which was one of the largest and most successful importers of footwear from the Far East. He had made a lot of money buying the sports brand Reebok for US$77,500 in 1981 which he sold ten years later for $770m. He was an astute businessman and, apart from Reebok, had built an impressive group with offices across the Far East. His business was partly owning and licensing sports brands as well as (like Browning, but on a much larger scale) supplying retailers with private label footwear and accessories.

I went to Pentland's award-winning headquarters in Finchley, north London. An amazing building constructed around an eco-friendly lake with large open workspaces with lots of natural light, very different from my small dark office in Maida Vale. Stephen, a friendly, studious-looking man with a full head of greying hair was welcoming and encouraging. He suggested that he provide all the support functions such as shipping and bookkeeping to leave me free to focus on designing, dealing with the factories and selling. Pentland would charge for these services at a reasonable rate. He would buy 51 per cent of the share capital for the net asset value on 30th June 1988 with a payment of £20,000 on signing the agreement and the balance fourteen days after 30th June 1988. It was attractive from the point of view of being part of a large successful group and allowing me to focus on growing the business, but I was less sure about parting with such a large part of the equity. I was also convinced I could do the operations and services a lot more cheaply than he was proposing. I thanked Stephen for the offer and said I would consider it and get back to him in a couple of weeks.

I never really had a mentor. Of course, my father and Len had been valuable role models, and I had learnt a lot from them about footwear, how to run a factory and manage people. But there was no one specifically that I went to for advice. However, a man whose commercial nous I respected was Monty Sumray, who was a good friend of my father. He had built a successful footwear empire which included Fiona Footwear, one of the largest suppliers of footwear to Marks & Spencer. I asked to meet him for his advice. He was adamant. Do not sell 51 per cent of your company. You will lose control. You've got to try and get through these difficult first years by giving away as few shares as possible. He introduced me to an Israeli bank called First International Bank of Israel (FIBI) where he was a non-executive director and arranged for a letter of credit facility to be made available. His advice was invaluable. I have no doubt that in the future I would have greatly resented selling 51 per cent of the company for a modest sum when the business was successful and making good profits. I got back to Stephen Rubin, thanked him and told him that I didn't want to sell a majority of the company at this early stage.

About the same time, I was having lunch with a good friend called Mark Sebba, who became the successful CEO of the premium online retailer Net a Porter, where he was greatly admired, not only for his management skills but also for his modest character and his slightly old-fashioned demeanour. I had known Mark a long time and had driven through Italy with him when we were in our early twenties. He was great company.

He liked good food and wine and had an excellent sense of humour. He was a heavy-set man with a distinctive, rather posh, accent. Mark suggested I speak to Gerald Newton who owned an investment company called Jermyn Investment PLC as he was looking at investing in small businesses with good growth potential.

I met Gerald, a tall man with a slight stoop. He was a quiet, understated character who had been successful in the property market. He was a good listener and after a couple of meetings when I explained my business plan and aspirations for Browning, he made me an offer to buy 29 per cent of the shares for £40,000 and put in a subordinated loan of £100,000. I was to use my remaining capital to subscribe for the remaining 71 per cent of the shares for £50,000 and provide a loan of £90,000. I was to receive a salary of £30,000 (worth about £80,000 today). The offer was very similar to that from 3i but a lot less onerous in terms of the requirements to meet targets and provide information.

In February 1988 I signed the agreement with Gerald. Jermyn Investment became my first, and only, shareholder. It remained a shareholder for five years when I bought back its shares for £300,000 and repaid the loan. Gerald was a passive investor. He let me get on with running the company without any interference. It suited me, but I did miss some of the encouragement and advice that an investor could have given. The only time he got involved was when we moved offices. He came with me to look at new premises. Property was his expertise. By this time, I had taken over the offices at 37a Maida Vale when the woman who ran the secretarial office moved out. We had outgrown that space and needed new offices and a showroom. We decided to move to a unit, which was half a floor, at Hatton Studios at 9 Hatton Street, five minutes' walk from my house. It was an old Spitfire factory that had been converted to studios by the architect Sir Terry Farrell, who had his offices on a floor of the building. We stayed in Hatton Street for 30 years.

With the new investment in the company, HSBC increased its letter of credit facility and together with the credit line from FIBI I was well placed to build Browning Enterprises Limited.

11. Far East travels

Having secured the necessary funding to grow the business I spent the next five years travelling up and down the M1 motorway to see customers in Leicester, Bradford and Manchester, showing them the ranges and flying to the Far East to get their samples made, check the quality, visit factories and put into work further development. It was the period in my life when I worked the hardest and travelled the most, although I tended to keep the foreign trips as short as possible so I could spend time with my family.

In those early days I enjoyed the travel. Being away from the office gave me an opportunity to think about the business in a more detached way. Not having to carry out the routine tasks, which took up most of the day, meant I had no distractions. Some of my best ideas were on the plane or waking up in a hotel room in the middle of the night. I would jot down my thoughts on the pad next to my bed. When I looked at them in the morning, those that were legible were a lot less interesting than they seemed when I wrote them, although there were one or two gems. At first, jet lag was a problem. Once I was awake my mind was racing, and it wasn't easy to get back to sleep. I found the solution was to try not to sleep too much on the plane, and to have a couple of glasses of red wine and a massage in the hotel before I went to bed. It didn't always work. The frustration of lying in bed, willing myself, unsuccessfully, to fall asleep was one of the least attractive aspects of going to the Far East, although the more I travelled the less of a problem it became.

Meeting the people and experiencing the varied cultures in the countries I visited was both enjoyable and insightful. The way of life was very different from the UK. These countries were going through a major transformation, from agrarian to industrialised societies. There was only a basic welfare state which meant that people had to fend for themselves and work hard to provide for their families. Food played a large part in their lives. Being a keen cook and food lover, trying the different cuisines and discussing food with the people I met was a special treat. Food is a universal subject that most of the people I met wanted to talk about. They were very proud of their local cooking. I was lucky to have some outstandingly good meals, mainly in modest family-owned restaurants. All the ingredients were locally sourced and bought in the neighbourhood market. Unfortunately, I didn't have the time to explore these countries as much as I would have liked as I was keen to get home to my wife and children. I had limited opportunities to see the tourist attractions. I spent most of the time either in air-conditioned offices or driving through the least attractive parts of the country on bumpy roads to the industrial areas where most of the factories were situated.

My regular trips to the Far East were initially to Taiwan and later to China, the Philippines, Indonesia, Thailand and Vietnam. As I wanted to keep costs as low as possible, for the first five years I flew economy class on Cathay Pacific to Hong Kong. I always stayed at the Regent Hotel, although unfortunately I wasn't upgraded to a suite. Perhaps they realised that I wasn't Stephen Rubin. The hotel was always a buzz of activity with a constant flow of people arriving and leaving. At the weekend there was often a wedding in the banqueting suite of the hotel.

They gave a good insight into how the wealthy lived in Hong Kong. The sweeping staircase, leading to the ballroom, was draped in magnificent floral displays. The guests arrived in chauffeur-driven Rolls-Royce and Mercedes limousines looking immaculate. They were dressed in designer eveningwear and were wearing trophy-sized diamond jewellery. Unlike the UK there were no concerns about flaunting your wealth in Hong Kong. If you had been successful, you were proud to show it off. In Hong Kong, everyone aspired to living a wealthy lifestyle.

In Taiwan, after my usual visit to Sunny in Taipei I would get the train or bus to Taichung to meet Janna Chen. Travelling in Taiwan was easy and efficient with both buses and trains leaving and arriving on time. This efficiency and reliability summed up Taiwan. Although it lacked character, and had a sense of impermanence, this relatively new country was a great place to do business. They wanted to please their customers.

I sometimes got a taxi from the airport to Taichung. On one occasion, when there was a torrential downpour, we were driving down the motorway when the car suddenly lurched to the left. Fortunately, we managed to avoid the other cars and came to a halt on the hard shoulder. I had been dozing after the long flight and was rudely awakened. The driver, a man in his late sixties, let out a string of what I assumed were Chinese expletives. We sat there for a few minutes, relieved that the car was in one piece. The driver got out of the car to review the damage. We had a flat tyre. After several minutes it was clear that he was struggling to change the tyre. I realised that I would have to get out to help him if we were to make it to Taichung before nightfall. The next hour was spent changing the tyre. We got completely soaked from the rain and the spray from passing cars. I have never been so wet. Not only was it raining but it was cold as well. I spent the rest of the journey huddled in the corner of the taxi, shivering even though the heating was full on. It was a relief when we finally arrived at the National Hotel. The taxi driver was clearly grateful for my help. He kept on bowing and muttering *"Xie Xie"*, thank you in Mandarin.

Having left the Japanese trading company, Sinco, Janna had set up her own company, Maxgreat, with her right-hand man Mike Wu. She bought a house in a quiet neighbourhood of Taichung with a pond in the front garden with koi carp. This was good feng shui, the ancient Chinese art of arranging buildings to achieve harmony and balance. The water in front of the house was meant to give you good luck in your financial dealings, which largely, through hard work, and maybe the pond, she achieved. Janna was possibly the most driven, hard-working person I knew. She had this infectious energy and enthusiasm for improving processes and was constantly striving to make better shoes and deliver them faster.

Sunny made the best plain leather courts which meant that I had to develop other styles with Maxgreat. I also had to ensure the styling was suitable for the UK market. Janna's main markets were Japan, Australia and Canada. She had no exposure to the idiosyncrasies of the UK. We developed a range of leather sandals and fabric court shoes that had some special styling details. I remember spending several days with Janna trying to produce a sprayed stacked heel. A stacked heel is a heel covered in individual strips of leather and stacking them in layers on top of each other to give the desired striped effect. The stacked heel can have a polished, natural or painted finish. They make them very well in Brazil. As the leather stacked heel was too expensive, we wanted to produce the same effect by printing or spraying the heel. We visited several heel makers, paint shops and material suppliers to try and get the correct finish. It proved to be a lot more difficult than we anticipated. Finally, after many attempts, the heel came out as I wanted. Most people would have given up on this challenge, wondering why it was so important. Janna was determined to get it right. It was one of the reasons she was so good to work with. She understood the importance of executing the details precisely and not accepting a compromise.

I spent a large amount of my time developing new styles. The inspiration for this development came from different sources. The meetings with buyers were very helpful. Reviewing and critiquing samples, trying them on the fit model to ensure both the cut and the fit were good, getting their feedback on the styling, were all essential ingredients in producing the right sample. Some buyers were happy to tell you what was selling well in their stores and what features were important. Others were more reticent. Some gave you specific shoes to develop. Others relied on their suppliers, like me, to come up with the styling. The whole process of exchanging ideas, talking about the direction of fashion, colours, materials, constructions was crucial in arriving at successful styles.

Another source of new ideas was walking around the stores in the many countries I visited, both in Europe and the rest of the world. Each town has a different slant on fashion. Amsterdam is more casual, Milan is dressy, Los Angeles is sporty, London is young, Paris is sophisticated. One of the unfortunate trends over the past ten years has been the disappearance of the smaller independent stores and the dominance of the major luxury brands, to the extent that you now see the same styles in every city around the world. This is a shame and leads to a homogeneous global range that becomes repetitive. The resources of the luxury brands means that they acquire the best stores, spend a fortune on marketing (their adverts dominating the front pages of *Vogue* and the other fashion magazines), have collaborations with the most famous celebrities, which gives them huge power in the market and makes it difficult for smaller brands to compete. Maybe that is a reason I enjoyed visiting Japan, which has its own culture and was less influenced by global trends. You could often find something different and unusual in the Tokyo stores, although Japanese fashion is more minimalist and sometimes less relevant for the mainstream UK market.

Street fashion was an important source of ideas. Seeing what people were wearing walking along the streets, in restaurants and clubs was often revealing. What pop and film stars were wearing mattered. Some fashion trends just emerged, caught the imagination and were picked up by the fashion influencers. Sometimes it was developed by a new brand, sometimes it was a vintage style. What made it a winning design worn by so many people was often a mystery. One thing was true: there was seldom something completely new. Most styles were reworks of something that has gone before. Fashion is now much more immediate and eclectic. Seasons have become less important as the quest for newness, stimulated by social media, has meant that brands are offering their creations when they are developed rather than on a seasonal basis.

Putting together the different ideas into a cohesive range of footwear was one of the most enjoyable aspects of the job. Armed with my photos, samples, tears from magazines, cutting of materials and my rudimentary sketches, I spent hours with the designers and pattern cutters developing the range, amending patterns, eliminating styles until we arrived at what I felt was a compelling offer. The hope was that the buyers would like it too. Fortunately, in many cases they did, although some of my favourite styles got the thumbs down. I did get carried away and ended up developing too many styles. After several complaints from the factories, I became more selective in what went into development.

My first large order for Maxgreat was from a footwear retailer called Faith Shoes. Faith was a privately owned company with about 50 shops and concessions in Top Shop, the go-to young fashion retailer of the 1990s. It was owned by the Faith family, started by the father, Sigi, and run by the son, Jonathan. Faith had a reputation for young fashion shoes. It was a well-run, successful company. Jonathan was strong on the numbers side of the business and had developed his own bespoke stock control system that was ahead of its time. He was an astute businessman who was brave and committed large quantities to styles he really believed in. He became a very observant Jew and tried, unsuccessfully, to get me to engage more in my religion. He was a large man with dark curly hair and a soft voice which was sometimes difficult to understand.

Court shoes were still the core of most women's footwear ranges. Sunny made the leather courts, so we decided to make fabric courts at Maxgreat using different materials including grosgrain, a fabric with distinctive ribbed lines. Jonathan loved the grosgrain court. He thought it would be a great vehicle for colour. He placed an order for 13 colours on three heel heights, in both a plain court and a slingback, in total 78 different options. The final order was for 60,000 pairs, a very large order and, to my mind, a risky one. I also doubted whether there would be demand for colours like purple, pistachio, mustard and yellow. I remember being in the office one evening as the orders came through slowly on the fax machine from Faith. This was in 1987, before the days of emails and printers. Page after page chugged slowly out of the machine, all 78 pages of orders. This was going to be an administrative nightmare and a weekend's work processing them to send off to Maxgreat.

Maxgreat placed the orders in three factories, one for each construction. Janna gave the factories initials, FCF, BT and BC. I flew to Taiwan to inspect the shoes. They looked good. Fabric shoes are difficult to make as they easily lose their shape in the lasting process and any mark or blemish is hard to remove without damaging the material. The shoes were shipped and arrived at Faith's warehouse. A few days later I got a call from Jonathan. There was a problem. I had a feeling of déjà vu. On some of the shoes the uppers had come away from the lining and had produced a bubble effect. This must have happened during shipping as the shoes I inspected were perfect. Just when I thought that quality issues were behind me, I was now faced with the possible return of 60,000 pairs of fabric courts in a variety of exotic colours. I pleaded with Jonathan to send the shoes to the shops to see if they would sell. He reluctantly agreed. If it had been any other customer, they would have refused and insisted I take the shoes back. The shoes were sent to the stores where they were priced at £14.99. Looking in his stores I was struck by the impressive visual impact of having 13 colours lined up on the display shelves. It was certainly a confident statement. Next week I had a call from Jonathan. The shoes were flying off the shelves. He would keep the stock and place a repeat on the bestselling colours. I breathed a sigh of relief. This was another example of the challenges of making shoes. Nobody anticipated that the uppers would become delaminated in shipping but the combination of the heat and humidity in the container had done just that. We hadn't shipped fabric shoes before. This was a lesson in trying to anticipate the unexpected, especially when manufacturing a new product. Janna remained supportive during this episode. She was very focused on quality. On the wall behind her desk was a sign saying: "The bitterness of poor quality remains long after the sweetness of low price is forgotten."

The business with Sunny and Maxgreat grew. With the plain courts I had carved out a sizeable niche in the footwear market. I also developed a wider range of shoes and sandals to complement the courts. Both Sunny and Maxgreat had grasped the cut of the patterns and the finer details that made the samples look good and attractive to UK retailers. My frequent trips to review and correct samples were paying off.

During the early 1990s the Taiwanese computer and electronics industries started to grow, and the cost of labour increased. Much as in northern Europe 10 years earlier, making shoes in Taiwan was becoming too expensive. Sunny and Maxgreat started to look for cheaper manufacturing sources. We visited the Philippines where one of Janna's factories, FCF, was producing shoes. Manila was a fascinating place, very different from its neighbouring countries because of its strong Catholic and Spanish heritage. The jeepneys, small privately owned buses that crammed the roads, were artistic and colourful, often with strong religious messages on the side or rear. They gave a unique atmosphere to the city, as well as a lot of CO_2 emissions. Unfortunately, the Philippines wasn't going to be a source of our footwear. The labour costs were too high and the prices uncompetitive.

The next country we looked at was Indonesia. We started making some shoes in both Bandung, near the capital Jakarta, and Surabaya, about an hour's flight to the east on the island of Java. The journey to the factories was particularly uncomfortable, being a long drive in an old jeep with poor suspension across dirt tracks with numerous potholes. Many of the shoe factories were started by businessmen of Chinese origin. The financial and business acumen of the Chinese had made them a powerful force in the Indonesian economy. The quality of the shoes was good and the prices competitive. However, the better communications and lower labour costs made China, which was growing in strength economically and politically, a more attractive destination. Towards the end of the 1990s there was strong anti-Chinese sentiment amongst native Indonesians which culminated in major riots in 1998 when Chinese-owned businesses and homes were attacked. Like the pogroms experienced by my grandfather in Russia, the riots were triggered by economic problems and food shortages. The Chinese became the scapegoats for the country's troubles and many of them left Indonesia.

Sunny had opened a factory in Thailand, near the capital Bangkok, making a major investment in the building and machinery. A group of Taiwanese managers and technicians was sent to train the local workforce and run the factory. It proved to be a failure and after two years the factory closed. The main reason it failed was the inability of the Taiwanese to manage the Thai workforce. The Taiwanese were used to a strict disciplinary style of management. Their domestic workforce was submissive and did what they were told. The Thais were more independent and free-spirited. They didn't accept the imposition of an authoritarian management style. It was a good example of having to adapt management style to the local population.

Sunny successfully relocated the factory to China where the workforce was more compliant and the costs lower. China became the dominant source of footwear. It had all the right ingredients. The proximity to Taiwan, Hong Kong and Korea, a large and low-cost labour force and the Chinese government encouraging inward investment made it the go-to country for footwear production. In 1988, I made my first trip to China, with Janna and Mike. The main entry point was via Hong Kong crossing the bridge at the Lo Wu in the New Territories on foot. Joining the crowds running across the bridge to the border point in Shenzhen I felt like a refugee pulled along in a stream of people. Everyone wanted to get the customs post first to avoid the queue. When, after an hour's wait, we eventually had all our papers checked and got through customs, a car took us to the factory. The roads were in poor condition and there were very few cars. The factory was an old multistory building in an industrial estate about two hours' drive from Shenzhen. A large automatic gate barred entry. A smartly dressed uniformed guard checked our identity. Once he was satisfied that we were bona fide visitors he opened the gate, stood smartly to attention and saluted as we drove into the yard. Outside the gate was a crowd of about 50 young people. These were economic migrants from the north of China looking for work. In the 1980s China was still predominantly an agricultural country and the wages that could be earned in the newly industrialised Pearl River Delta were a strong incentive to move south.

We met the owner who gave us a tour of the factory. It was a replica of the factories in Taiwan although larger because there was a dormitory where the workers slept and a large kitchen and dining area where meals were eaten. The conditions were basic but clean and functional. There was a large cadre of Taiwanese managers assessing the Chinese recruits' hand/eye coordination, dismissing those who didn't meet their standards and training those that passed the test. There were three production lines, each one making a different style. Although production was slow the quality was acceptable. Several Taiwanese quality controllers examined the production at different stages to ensure quality standards. The owner said the priority was making good quality shoes. Production speed would increase slowly as the operatives gained confidence. He said generally the Chinese were conscientious and hard-working. The only trouble had been when the workers had briefly gone on strike because they were unhappy with the quality of the food. A new cook had been employed and this had solved the problem. Now everyone was happy.

I was always impressed with not only how hard the Taiwanese worked but also how mobile they were with teams of workers leaving their families in Taiwan to set up production units in China and later Vietnam, Cambodia and Bangladesh. They often worked on a rotational basis with managers going home after a month's shift and returning after a week's break. As footwear production in Taiwan continued to fall more, they travelled to other Asian countries more often, as did I. And they increasingly went to mainland China and Vietnam.

12 – China opens up

During the end of the 1980s and the beginning of 1990s China became our main source of footwear. The closeness, both physically and culturally, to Taiwan made it the preferred place to open factories. We retained some production in Taiwan. However, this was steadily reducing as costs increased and Taiwan focused on the technology and engineering industries. Many of the factory owners moved their production from Taiwan to China with the city of Dongguan in the Pearl River Delta being the preferred location for dressy shoes. Fujian province on the south-eastern coast was the main centre for sports and Wenzhou for men's footwear. The Hong Kong factories also closed their local production units and moved north to Shenzhen and Dongguan.

Janna relocated a lot of her team to China. Initially she worked out of one of her factories until she bought an old school building with a running track in the suburbs of Dongguan where she set up an impressive operation, both as her head office, sample making facility and product testing centre. My production was gradually moving away from Sunny. It finally ended when I got a call from Jason Yang to meet him at a hotel next to Heathrow. His message was simple. He wanted to terminate our relationship and deal with an ex-buyer of the BSC who had started his own import company. He must have felt that having a close relationship with an ex-BSC buyer would improve his prospects with them. Given that I had been working with Sunny for almost ten years and had successfully built his business in the UK and supplied him with a lot of knowledge on the UK market, I was disappointed, especially as he wasn't offering a notice period beyond the fulfilment of my current orders. As there was no contractual arrangement, I didn't have any grounds to object, apart from expressing my disappointment. I wasn't in favour of contractual relationships with suppliers although in this case it could have given me greater security. On the plus side I was free to move factories and suppliers more easily if the partnership wasn't working satisfactorily. I have always been suspicious of legal agreements in these circumstances.

Personal relationships are the most important ingredient to a successful partnership. The fact that I am still dealing with the same suppliers more than 30 years later supports that argument. The split with Sunny gave Janna an opportunity to take over our substantial leather court shoe business which she did in her typically professional and efficient way, although getting the right softness and finish of leather took a little time.

I continued to fly to Hong Kong and stay at the Regent Hotel until it miraculously changed in 2001 to the Intercontinental. It was like magic. I was there in June of that year. One day it was the Regent and the next day it was the Intercontinental. Everything was rebranded overnight. Every single item from the ashtrays to the towels, the uniforms to the signage changed. I tried hard to find something they had missed but was unsuccessful. In other respects, the hotel didn't change. I continued to enjoy the stunning view of Hong Kong Island and be mesmerised by the vessels going in and out of the harbour. In 2020 the hotel reverted to the Regent when the ownership changed. It reopened in November 2023 after a major refurbishment.

I had found a handbag supplier in Hong Kong called Minglo that made beautiful dress bags. Some of them were works of art with intricate diamante and jewelled details. It was run by the Ng family. I never really succeeded in pronouncing their name correctly. Their offices and showroom were in an industrial building in Kowloon, on the way to the old airport at Kai Tak. The building had a particularly temperamental lift, and it was always a relief when I reached their offices on the eighth floor. I dealt with the daughter, Minnie, a petite young lady who was mature beyond her years. She was very efficient and a pleasure to deal with. Mr Ng, I never got to know his first name, was a quiet man with very thick glasses and a full head of greying black hair. As his English wasn't good, he left Minnie to deal with customers while he and his sons looked after the production.

They were a hospitable family and took me to dinner when I was in Hong Kong. I was surprised when Mr Ng took me to a karaoke club. We entered this dark cavernous room, and Mr Ng was greeted warmly by the manager. He was clearly a regular. After a couple of drinks, he got up on the stage and sang this slow, emotionally charged, love song, at least that's what it sounded like. He had a pleasant voice, but it was the feeling that he put into it that was impressive. I saw Mr Ng in a completely new light. No longer the quiet, almost studious, executive but a soulful singer of Chinese love songs.

Another evening, we went to the fish market in the New Territories. We bought a large live grouper fish that was given to us in a plastic bag where it was still moving about vigorously. We took it to a local restaurant where we handed over the fish and Mr Ng gave instructions for its preparation. It was one of the freshest fish I have eaten. It tasted good but surprisingly it was a bit tough. Some fish need to rest to get the best texture. The biggest fish restaurant I saw was in China near Dongguan. On entering the restaurant, customers were faced with a display counter that must have been the width of a football pitch with a huge variety of fish including a cage that housed some small, or dwarf, crocodiles and another containing turtles. Every type of crustacean was on display as well as eels, catfish, sturgeon, abalone and the famous dragon fish, the ultimate symbol of wealth in China. We passed on the turtles and caiman and ate a delicious, unexotic, white fish in a black bean sauce.

Every year the production from China was improving and our business expanded. While our business with Maxgreat was growing we also found other suppliers outside Guangdong that specialised in more casual products such as fashion trainers and cheap sandals that appealed to the supermarkets that were increasingly growing their footwear business. The factories were in a city called Ningbo, just south of Shanghai. We also started a men's formal footwear business from a factory in Wenzhou, a large city on the east coast. We found a number of these factories at trade fairs such as Garda and Dusseldorf, as well as receiving direct approaches from the factories themselves which were becoming increasingly outward looking.

Visiting new factories was an essential element of building Browning. A bit like looking for a new home, you instantly got a feel as to whether it was right or wrong. The most important factor was meeting the factory owner. Even if he didn't speak English, you got a sense of his commitment and pride in his production. Walking round the factory gave me a clear picture of whether I felt confident it would produce shoes of the desired quality. In some I wasn't confident that would be the case, either because the building was in poor repair, the shoes on the production line were poor quality, or something less tangible, just a sense that workers were unhappy and were badly managed. I have always believed in visiting the factory and meeting the management. You also get an interesting insight in what they are making for other customers. Many buyers limit themselves to going to the showroom, which is a mistake. Controlling the quality of new factories was critically important as the risk attached to a new factory where we had no trading history, as I had found to my cost in the past, was significant. At the end of the 1980s we hired a quality controller, Jason, whose main role was visiting these factories to check production to ensure that our quality standards were being met and that the factory met our increasingly stringent ethical standards.

Ensuring that labour and factory standards were acceptable was a key aspect of our due diligence when we dealt with a new factory. We had a checklist that covered areas like the age of the workforce, the hours worked, the amenities like toilets, kitchen and living quarters, and the general working conditions. We carried out a fire risk assessment as losing a factory and its production would be very costly. Reviewing the machinery and their operations was also important. Were the machines safe, did the factory have the right extractor fans to disperse the fumes from the adhesives? Most of these checks are now carried out by specialist companies like Intertek although visiting the factory is still essential. Based on experience a good quality controller can get a sense of whether a factory is one that we want to deal with.

With Maxgreat we grew a successful children's business which we sold to Next and Marks & Spencer. This was run by Mandy Ellis who joined our team. She was the ultimate specialist in kids' shoes and built this business into an important part of our sales. She is still a major supplier to the trade some 30 years later. The design strategy for kids (mainly girls) was based on "mini-mes". Taking appropriate women's styles and grading them down to kids' sizes. Girls wanted to look like their mums and often Clarks and Start-Rite, the standard brands for kids, were unexciting and lacked that element of fashion that kids craved. The notion that if you didn't buy Clarks or Start-Rite for your kids you would do irreparable damage to their feet was waning. The days of queuing at John Lewis, waiting for a ticket, having your feet measured on a special device and sometimes an X-ray fitting machine, were coming to an end as store groups released the potential for selling kids' shoes.

We were unsuccessful in building a range of sports shoes despite a few attempts. Our strength was fashion. It required a different talent to build a sports shoe. It was more technical, and the market was dominated by brands such as Nike and Adidas who were placing huge orders. These brands had a large product development team that were making important strides in improving comfort and performance, such as the air technology for sole units. We just couldn't compete. The unbranded sports footwear market was highly competitive with slim margins that made it unattractive. With hindsight given the huge success of new sports brands such as Hoka and On I often wonder if that was the right decision.

Getting around China was also getting much easier. Road and rail connections were rapidly improving. I was still nervous flying on the local Chinese airlines as their safety record and service proposition at that time wasn't reliable. But one of the most frightening experiences was taking the train from Guangzhou back to Hong Kong. I was with Mike Wu. We were coming back from a factory visit and were late for our train. The station was packed with commuters. It was difficult to move. We had no idea which platform our train was leaving from. The signs were in Chinese, and no one could understand Mike's Taiwanese accent. We rushed around frantically trying to find the right platform. It was extremely hot, and we seemed to be going round in circles. Mike was in a panic and was sweating profusely. I wasn't optimistic that we would find the train. After 30 minutes of getting nowhere we were about to give up when an elderly Chinese man saw we were struggling. He spoke perfect English and escorted us to the right platform where we just made the train. It was interesting comparing the station experience in China with that in Delhi in India. In both stations you were overwhelmed by the sheer number of people. In India it was chaotic but there were lots of smiling, friendly faces. In China everyone avoided eye contact and despite the huge number of people you felt completely isolated and helpless.

In 1994 the European Union placed quotas on Chinese leather shoes. This was a huge blow to our business. The quotas were meant to protect the producers in southern Europe, in particular

Italy, Spain and Portugal, who were struggling to compete with Chinese production. Most of the footwear factories in northern Europe, like my old factory, had closed as their costs were too high. Unless they produced a distinctive product that couldn't be replicated in China, such as goodyear welted men's shoes, they found it impossible to compete with Chinese production. The southern European factories were experiencing the same challenges. The factors that enabled factories to survive were quality, quantity and speed. On quality the factories in China couldn't match the look and feel of the best manufacturers in Italy, Spain and Portugal. There was a degree of craftmanship, often built over generations, which together with the quality of the leather and components, gave these factories a unique product that couldn't be copied. The Chinese factories also relied on big quantities. They were not set up for smaller orders. This meant that smaller factories in Europe, which were more flexible, had a role to play. Finally, the lead times from China of up to five or six months meant not only a large stock commitment for the retailer, with the risks that this posed, but also the scope for placing repeat orders on successful styles was greatly limited. Retailers in Europe could get their orders in a few weeks rather than months, so although the intake margin was higher from China after allowing for stock markdowns and greater flexibility the difference on the achieved margin was not so great. Over the years Chinese factories have become more flexible although the long lead times remain a problem for a fashion product. The cost of air freighting footwear to the UK is prohibitive, unlike clothing which is lighter and less voluminous.

The quotas lasted from 1994 until 2004 when China joined the World Trade Organisation and the quotas were relaxed. There was no strong evidence that the Chinese government was subsidising the footwear industry. It was more a sop to the Italian, Spanish and Portuguese governments. In the end it didn't greatly help these factories as production continued to move from China to other Southeast Asian countries and the number of European factories continued to close. Although the quotas were for the whole of the EU there were big differences in how they were enforced. Whereas other countries took a more relaxed view on applying the quotas, the UK set the standard of ensuring that no shoes got past customs unless every last piece of documentation was correctly completed. We are a very law-abiding country which is laudable, although it can be frustrating. In the end we brought shoes into the UK via Holland or Italy as their customs were more relaxed and quota was easier to obtain. Because we were in the EU there was free movement of goods between EU countries.

13. Production moves to Vietnam

My first visit to Vietnam in 1993 didn't start well. I was meeting Janna to get a better understanding of the local production. I flew into Ho Chi Minh City (formerly Saigon) in the south of the country where most of the factories were located. The customs officials at the airport were aggressive and refused me entry because my visa wasn't completed correctly. They waved me away from passport control. I reviewed the visa. It looked to me like it was completed fully and accurately. It wasn't until a German man, who saw what was happening, explained that I needed to give the official money, that the problem was solved. I went back to the official at passport control who was dressed in a green military uniform, with a gun hanging from his waist, and handed him my passport, the visa and a ten dollar note folded into the pages of the passport. He smiled, stamped my passport, ignored the visa, and waved me through.

Given the challenges of importing leather shoes from China from the mid-1990s onwards, we looked for other sources to make our shoes. After considering several countries Vietnam was chosen as the preferred country while the quotas were enforced. It had a large, young workforce who were efficient and talented. As the country moved away from an agrarian economy the opportunity of working in a factory and earning a better wage was attractive. The Vietnamese government, although communist, was keen to attract inward investment and encouraged Taiwanese factories to relocate to Vietnam.

Some Taiwanese joint venture factories were already operating in Vietnam and these added production lines to increase their output to meet the increased demand. Janna worked with her factories in China to set up a manufacturing facility in Vietnam. The speed of the move to Vietnam was impressive and was a testament to the efficiency and mobility of the Taiwanese. There were delays as the factories geared up for the increased demand and the move wasn't seamless, but overall, it was successful. Our sales were adversely affected in 1994/5 due to the disruption. The cost of the product was similar to China, although I suspect that the manufacturers had been making a large profit from their Chinese operations. There was also a delay of about two weeks in the lead times as most of the components had to come from China. Overall, it was a hiccup we could have done without. It is frustrating when your performance is affected by factors outside your control, in particular government action which is poorly thought through and without proper consultation with the companies it will affect.

I enjoyed going to Vietnam. After China the country had a relaxed and welcoming atmosphere. The people were friendly and polite. The food was exceptional, like Thai, but with a hint of French cuisine. Vietnam was a French colony from 1894 to 1954 and there was still a Gallic influence in the food and architecture. Although the war between the communist North and the US-backed South ended in 1975 there were still remnants of old military planes and bunkers at the airfield at Ho Chi Minh City. After 20 years of a bloody war there was no lingering sense of bitterness or resentment against foreigners. Ho Chi Minh City was like Bangkok many years earlier. There were only a few cars. Most of the locals got around on bicycles and motorbikes. Often the whole family were perched on a motorbike. The most I counted was a family of six people sandwiched together on a motorbike. Tall thin men had huge bundles of goods balancing on their bicycles as they rode unsteadily through the streets to deliver their wares – very different from the UPS and DHL vans used today.

Most of our factories were one or two hours' drive from the centre of the city. The roads were bumpy and the suspension of the people carriers was poor. We drove through small suburbs of the city which had open fronted workshops, food stalls and young men sitting cross-legged on the floor smoking. Children were going to school smartly dressed in white shirts, navy and black skirts or trousers and red scarves. I noticed all over Southeast Asia, despite the poverty, schoolchildren were well dressed in distinctive uniforms.

The factories were much like those in China except generally smaller. The factory owners greeted us warmly. They were not used to English people visiting their factories and proudly showed us around. There were a lot of quality control (QC) stations manned by Taiwanese managers as they were conscious of the need for quality standards to be kept high from this relatively new source. The workforce was made up of a mix of young men and women who seemed to be working enthusiastically although slower than those in China, as many were new recruits. The production looked good although there were several racks where the QC team had taken shoes off the production line to be reworked or rejected.

One of the larger factories was Strong Bunch (or SB as Janna referred to it). They were making a lot of shoes for Clarks. We placed a high platform sandal in the factory. The usual quantity for a new construction was 10,000 pairs. I had persuaded Janna and the factory to adopt the construction for 4,000 pairs. It proved to be one of our best sellers. We sold over 250,000 pairs.

As this was the first high-heeled platform sandal we had placed in the Far East we spent a lot of time reviewing the construction. The attachment of the heel and platform was critical as there would be a lot of pressure on these components. If not attached correctly, they could easily fail in wear. The factory owner, Steve, relished this challenge, and the final product was excellent, and importantly the failure rate was very low.

Another factory was Beauty (BT) which we worked with for many years. The factory always looked a bit disorganised and cramped but the quality of the production was reliably good. The owners liked to chat about football and golf. There was a set of golf clubs in the corner of the office although I don't think there was a golf course in the country. They must have practised in the field behind the factory. Like a lot of the men I met, they were passionate supporters of Manchester United. David Beckham was their hero.

Mr Pang was an interesting character. He owned a factory whose name I have forgotten, most likely because I always thought of it as Mr Pang's factory. He was a short round man with a shiny bald head. Before you could discuss shoes there was a tea-making ritual which lasted about half an hour. Mr Pang would boil the water, pour it over the teapot which was on a large red porcelain tray with the six small teacups arranged around it. This left a large puddle of water on the tray. He put the Oolong tea in the pot, swirled the water around the teapot, waited five minutes while it brewed and then poured it into the red cups, throwing away the first two cups and finally pouring his guests their cups. This operation was usually done in silence so that Mr Pang could concentrate fully on the tea making. The resulting tea was good although I have always preferred strong Assam tea to Chinese green tea. Fortunately, Mr Pang's shoes were made as conscientiously as his tea and were generally of excellent quality.

The most sociable of the owners was Jimmy who spoke excellent English and was a mine of information. He liked to have cocktails on the roof of the Rex Hotel in the centre of Ho Chi Minh, a popular gathering location for ex-pats. During the war it was the home of the US Information Service and a famous haunt of US officers. From there you had a good view over the city. On a balmy night, with the swallows flying overhead, you sipped the local cocktail, the Pho cocktail, an unusual mixture of vodka, pho broth, lime juice, fish sauce and aromatic herbs and spices. To be honest I usually chose the local beer, Saigonese lager 333, as the combination of vodka and fish sauce was an acquired taste. Jimmy liked quizzes. He bombarded us with interesting facts. One question he asked was which country had the first subway system. After several wrong answers he put us out of our misery. The answer he gave was Buenos Aires. Some years later I discovered that his answer was wrong. It was London. Jimmy was entrepreneurial. He opened one of the first factories in Cambodia where labour rates and duties were low. After a couple of years, he stopped selling to us and became New Look's biggest footwear supplier and made a lot of money.

I continued to visit the factories in Vietnam and China and offices and shops in Hong Kong although from 2000 the visits became less frequent as I focused more on building the retail business. There were some memorable moments; some good, some not so good. I was in Hong Kong on June 30th 1997 when, in the torrential rain, the territory was handed over to China after 156 years of British rule. For some years there was no discernible change. Gradually China began to impose its laws on the territory leading to the elimination of democratic institutions, demonstrations, and the arrest of dissidents who criticised the communist regime.

I was in Taichung in 1999 when there was a large earthquake. The stairs of the building were shaking dangerously as I hurried to evacuate the hotel – a truly frightening experience. I was in Hong Kong during a monsoon when there was torrential rain and fierce winds. I was confined to my hotel room for 24 hours with the curtains closed. I did make the mistake of venturing outside but got completely soaked as soon I opened the front door. I quickly returned to my room.

There were many enjoyable moments. Janna had a memorable party to celebrate Maxgreat's tenth anniversary. There were lots of guests; factory owners, customers and friends. The running track was full of stalls offering many varieties of delicious food, there was a ceremonial dance with actors dressed as dragons and all sorts of entertainment. I had to give a speech. I think a lot of it was lost in translation as one of Janna's assistants attempted a simultaneous translation into Chinese.

We had a party in Taichung with the factory owners to celebrate ten years of Browning working with Maxgreat. There was an amazing amount of brandy drunk as everyone toasted each other with shouts of "Ganbei", a Chinese equivalent of "Cheers". The literal translation is "empty cup", which is used to encourage guests to finish their entire glass. Many glasses were emptied that night and sore heads felt the following morning.

14. Building a team

Browning had grown to a size where I needed to start building a team. My time was most valuably spent building ranges and selling. If I could delegate some of the administrative tasks it would free up my time for these more revenue-generating activities. One of the big lessons from those early years, and one that is obvious but often forgotten, is the need to get out there and meet with customers. As the founder of Tesco, Sir Jack Cohen, said, you can't do business sitting on your backside. There is a fine balance between stalking a customer and seeing them regularly, so you are top of mind when they are placing orders. You must have a reason to see them, such as showing a new design or getting a style they have bought confirmed. I have always found visiting frequently and building a relationship is crucial.

Persistence was certainly a factor in receiving my first orders from the British Shoe Corporation. Once the Dolcis buyer had bought the leather court shoe, it became a bestseller and from that date onwards I became an established supplier. As well as travelling to BSC, I was also visiting the other multiple retailers like Olivers, Stead & Simpson, Tandem and Lennards in Leicester,

Stylo Barratts in Bradford and Timpsons in Manchester. All have since closed.

The downside of spending so much time visiting customers was that there was a lot of time travelling, waiting in receptions to see buyers and generally hanging around. I got to know the M1 motorway very well during this period. During one visit to Stylo Barratts in Bradford I spent four hours driving to their offices only to be told when I got to there that the buyer was unwell and couldn't make the meeting. I then had a four-hour drive back. As this was before the use of mobile phones a lot of time was spent thinking and listening to the radio. During some of the trips I had to stop for a quick nap, as despite having all the windows open and singing along loudly to the music on the radio, I found my eyes slowly closing. I put on weight during this time as I devoured packets of crisps and snacks to keep busy.

I was getting bogged down preparing orders, requesting confirmation samples, chasing deliveries and arranging transportation. These were important tasks. Making sure that deliveries were on time and all the customers' complicated documentation was correctly completed were almost as important as delivering the shoes. Customers were quick to penalise you and issue a debit note if there was a small discrepancy in the paperwork. However, you needed to get the order in the first place and that was my top priority.

I was also getting behind with the accounts, which wasn't a good look for a chartered accountant and wasn't appreciated by my investor. Given the precarious state of my finances, it was essential that cash flow was tightly managed, invoices were sent out promptly and payments actively chased, and I was getting behind with these tasks. Ironically it has always been the area of business I have enjoyed the least and have been pleased to delegate to someone who is better suited to manage the finances. Having said that, my accountancy training did give me the ability to look at the big picture, read a set of accounts and understand where the challenges were likely to occur.

My first employee was Yvonne Overton, a woman in her thirties with curly blond hair, and a bohemian dress sense. She was confident and efficient and immediately relieved me of a lot of the administrative chores. My next employee was Rashmi Kothari, the bookkeeper. He was a quietly spoken Indian whose family had been forced to leave Uganda when Idi Amin, the President, expelled the country's Indian minority. Rashmi worked for me for over 30 years. What he lacked in dynamism he made up for in reliability, consistency and commitment. I can't remember him ever taking a day off for sick. His slowness could be frustrating, and he certainly found learning new technologies a challenge, but I always knew that what he did would be a hundred per cent accurate and on time.

My main hire was the sales director, David Brook. He was working at one of the large importers, ISA, before he joined me, so he had excellent contacts. He was a complex character. He had a natural charm that endeared him to customers and a strong work ethic. He was my number two and made an important contribution to our growth. He was a good person to bounce ideas off and shared my enthusiasm as we increased our sales and profits. He was a good travelling companion. On one occasion we were attending a trade fair in Bologna before driving to Florence to meet our agent. We were recommended a restaurant that specialised in spaghetti with lobster, one of my favourites. We arrived at the restaurant and there was a long queue. The owner kept on saying the table would be ready in five minutes. After waiting for 45 minutes, we lost patience and went to the restaurant next door, which was empty, where we ordered the spaghetti with lobster. I don't think I have ever felt so ill. The lobster must have been off. Apart from stomach cramps and a high temperature I was violently sick. David took control and drove us to Florence. I don't remember much about the journey except shivering uncontrollably. I staggered to the hotel room where I slept for 24 hours. Since that incident I have been very choosy where I eat my shellfish.

As time went on, David became more erratic and lost motivation. His attendance at the office became patchy. He eventually left and set up on his own. This was a key challenge of hiring entrepreneurially minded executives. The cost of becoming an agent for a factory, or an importer were low, especially if you were funded by the factory. As a result, a few of my most successful employees decided to leave and set up on their own.

I employed a talented designer, Philippa Poole. She added a new dimension to the range, especially on the dressier styles from Spain, which became an important element of our collection and the basis for our retail collection. This allowed me to prepare a more varied range and present it in a more professional and detailed way although I still took lots of photos (and still do) and scoured the fashion magazines for ideas and inspiration.

During the 1990s and early 2000s the business continued to grow rapidly and became more profitable. Browning became one of the largest designers and importers of unbranded footwear in the UK, supplying most of the important retailers and importing over 10m pairs of shoes a year. By 2005, our peak year, turnover had reached £60m and we were making a profit of £3.5m. We had a strong women's dress shoe business, and other categories, in particular children's shoes, were starting to perform. Our focus of adding value through design was paying off as the styles we supplied were often the retailers' best sellers. We had built a slick operation that made us the preferred supplier of many of our customers. Our business model was uncomplicated. We designed, we bought, and we sold. All the stock we bought was supported by back-to-back orders from the retailers, so the risk, unless there was a quality issue (which, fortunately, was now rarer) or late delivery (an ongoing issue), was low. I have always felt that there is a danger of overcomplicating business. If you have a great product (importantly with a point of difference) and you are totally committed to selling it, you have a good chance of being successful.

About half a dozen importers/wholesalers were our main competition, although some of the larger retailers, like the BSC, dealt directly with the factories or the factories' agents. John Wilkes and ISA were the largest and more established importers that had started importing from Europe in the late 1960s and early 1970s before expanding to the Far East and India. They had built a formidable reputation based on their long-standing relationships with both the factories and the retailers. There was Pentland, which was particularly strong on sports and casual brands. Our competition was some of the newer importers like John Westmacott and Keith Childs who were both based in Leicester and offered a more fashionable product. Keith Childs was a colourful character who had his showroom in his grand country estate in Leicester which was an added attraction to some of the buyers. Several niche importers specialised in specific countries or products. New competition emerged as European companies saw the attractions of the UK market and, as I had found, my more entrepreneurial employees decided that they would set up their own business or act as an agent to a foreign factory

To support the growth, I continued to build the team as we expanded from women's shoes into men's and kids' shoes. It was a strong team with some outstandingly talented and committed people. Helen Gilhooly was a merchandiser who became my PA. She was a feisty Irish lady who was super-fast and efficient. Susannah Huller, an ex-buyer from Barratts, had a great feel for shoes and was a dedicated and successful salesperson. She became our sales and product director. David Jameson was the most committed and hard working (and successful) salesperson I have known. Unfortunately, he left to join a competitor, although we remained on good terms. With hindsight I should have persuaded him to stay as his hunger for selling was exceptional. Duncan Miller, a personable and proud Scot, was our finance director, who became managing director in 2004. He had been a first-class finance director and was a sensitive and effective leader. As we grew, we became more structured and started to lose some of the entrepreneurial drive that had been such an important ingredient of our success.

From 2003/4 our margins were starting to be squeezed as new competitors entered the market who were leaner and faster than us. Many of them were small operations, with one or two employees. Some were agents of Chinese factories or trading companies. Both had a low-cost base which meant that they could afford to cut prices below ours. Increasingly our customers were driven by price and margin and were reluctant to pay a premium price for a superior service and product. The larger importers, such as us, were being squeezed as retailers bought directly from the factory or local supplier, or one of the smaller operations working on low margins. At the same time our overheads were growing fast. The ratio of overheads to sales, a crucial performance indicator, was getting too high, especially as margins were coming under pressure. Our main overhead was staff and by 2007 we were employing 99 people, 64 in the UK and 35 in Hong Kong and China, at a cost of over £4m. This was not sustainable.

Another factor that affected our margins was that we were making more and more samples, partly at the request of customers and partly because our design and sales team were being too extravagant in their demands. A salesperson always likes a big collection of styles to offer although the customer ends up only buying one or two styles. There was a further problem as the factories often charged for samples indirectly by increasing the price of a product, so the cost was hidden and our margin adversely affected. The freight cost of flying the samples to the UK was also high. The bottom line was that we couldn't achieve the margin that was required to maintain the level of service we were providing. Increasingly the customer was taking the service for granted.

Our customers were also changing. Specialist retailers were fast disappearing. Discounters, clothing and department stores were taking their place. Many of them treated footwear as a commodity. The buyer often came from the lingerie or childrenswear department, so had little understanding of the sector.

It was time for a change.

15. First attempt at retail

It was becoming more and more obvious that I would have to build a business where I had greater control. Cheaper competition from overseas had sunk my manufacturing business, London Lane. I had become an importer to try to stay ahead of the game. Now those foreign factories were starting to deal directly with UK retailers and I was getting squeezed again. The only thing to do was to start dealing directly with the customer myself, create a brand and become a retailer.

In 1991, I took my first steps towards becoming one. We were designing and making a range of stylish shoes in a small factory in Elda, near Alicante in Spain, called José Molina, which we were selling to some of the better grade high street retailers. I had been working with the factory for a couple of years. They made beautiful kid leather court shoes and sandals. The factory was owned jointly by José and his partner Fermin. José was a diminutive man with a bald head and round rimless glasses. He was an ideal partner; conscientious, hard-working and loyal. His shoes had a delicacy and elegance that was exceptional. The factory was a small unit, especially compared with those in the Far East. There was a conveyor system which transported the shoes from one operation to the next. There were about 20 workers making 150 pairs a day. Most of the workers were experienced shoemakers, and many of them multi-tasked, moving from lasting the shoes to attaching the soles and heels. Unlike the Far Eastern factories where the focus was on productivity, the priority at José Molina was ensuring that the quality was exceptional and consistent. Kid leather is a fine material with small skins, so it was essential that it was handled carefully from the cutting through to the lasting and finishing of the shoes.

There was a strong work ethic in these small factories. They were often open late into the evening to complete orders. The different mealtimes dictated the hours worked. Lunch didn't start until around 2.30pm and dinner around 10.30pm. If you arrived in a restaurant before 9pm you were often the only party eating. In those days road safety wasn't so strict, and we often left the restaurant having drunk beer, wine and a digestive, usually brandy de Jerez or whiskey. I often craved a siesta after lunch having consumed a large, boozy meal.

I visited the factory regularly with my new agent, Paco Mas. Paco was an engaging, friendly man who knew the industry well and had strong relationships with many of the local factories. He was about my age and height and had a scruffy beard which he fiddled with when he was thinking. We shared a love of food and wine. One of my favourite meals was when he took me to a restaurant in the hills above Alicante. It was a modest place on the edge of woods. It specialised in paella with *coneja* (rabbit) and *caracoles* (snails), not a dish I would immediately be attracted to, but it was delicious with a woody, rustic character, that I haven't tasted anywhere else. He ordered a red wine called Vega di Sicilia, a rich full-bodied wine from a small estate in the Ribera del Duero in Castilla in the centre of Spain. The combination of the paella and the wine was memorable. Like a lot of the local people I met in both Spain and Italy there was pride and interest in the regional identity and terroir. Hunting, shooting and fishing were strong weekend pursuits with hunting wild boar a particular favourite. For an urban creature like me this was a refreshing change. I was invited to join in these male-dominated activities but graciously declined. I had to get back to London and my family.

We regularly dined at La Serena in Elda. There was a large display of fish at the entrance of the restaurant that emitted an enticing salty smell of the sea. We ate a large plate of *langostinos de Santa Pola*, *cigalas* (crayfish) and *gambas* (prawns) followed by *merluza* (hake) with peppers, chorizo and paprika, a winning combination. Paco took great pleasure sucking the heads of the prawns to extract every last morsel. Santa Pola was the local seaside resort where many of the locals spent their weekends in the summer months. A low-key resort with no foreigners, no pubs, no fish and chips shops or burger joints, just great family-run restaurants and excellent tapas bars. Paco, who was always looking for a good deal, persuaded me to buy a flat in Santa Pola as an investment. Unfortunately, the timing wasn't good. Soon after Spain experienced one of its many property price crashes. The flat didn't prove to be one of my best investments.

Two retailers, Faith and Barratts, who bought our range of Spanish courts, suggested that we open a small concession in their flagship stores on Oxford Street. I liked the idea of offering this range with our own brand, rather than being a part of the Faith or Barratts range. It would give us our own identity and over time would reduce our reliance on the private label, unbranded Browning business. The name I decided upon was "Comme il faut", a French expression which means: "As it should be". At the time French expressions were in vogue, and I liked the concept of selling shoes that were *comme il faut*. The words, as spoken by French people, had a nice ring to them. It turned out not to be a great choice. Our customers found it difficult to say the word *faut* (in French it is pronounced "foe", the "t" is silent). It was becoming "comme il fault" or "communal fault" which was not the message I was trying to express. The concessions in both Faith and Barratts were not particularly successful.

The main problem was that our shoes were too expensive for the environment in which they were being sold, especially in the case of Barratts. I was encouraged by the feedback from customers which was overwhelmingly positive. I had spent my Saturdays serving in the concessions. It was valuable hearing their comments first hand. Customers liked the styling and the quality and felt that the shoes compared favourably with some of the luxury brands which sold for more than treble the price. I was excited because this was a strong foundation for building a successful niche retail brand.

I also learnt a lot about the fundamentals of retailing, in particular understanding how much stock to buy. One of the challenges of footwear is the sizes. On women's alone there are a minimum of seven sizes and if you buy half sizes as well that goes up to eleven or more. This compares with three or four sizes for clothing. Over the years this stock commitment has been one of the main difficulties of running a chain of footwear stores. So much capital is tied up in the stock. And if you buy the wrong stock, as can easily happen, you must discount it, which can be expensive and adversely affects your margin.

During the 1990s and 2000s, footwear retailing was undergoing fundamental changes, just like manufacturing. Specialist retailers, which had dominated the sale of footwear, were gradually losing their prominence. Increasingly shoes were being sold through department and clothing stores, discounters and supermarkets. From 2000, online retailers, such as Amazon, were also getting in on the act and taking market share from the specialists. At the same time sports trainers were becoming increasingly popular, not just for performance sports activities, but also as casual footwear. An industry that had been an important vertically integrated contributor to the UK economy was becoming fragmented and was going through a period of irreversible change.

Specialist retailers at the lower end of the market were the first to go. Lennards, part of Greater Universal Stores, with 260 stores closed in 1992. Timpson's was a loss-making retailer with 500 stores that was sold to George Olivers in 1987. (Timpson's has since reinvented itself as a successful shoe repair, key cutting and many other services company under the excellent leadership of John Timpson and his son.) George Oliver, founded in 1860, and in 1887 reported to be the largest footwear retailer in the world, became loss making and after closing many of their stores was sold in 2000 to Shoe Zone for £6m. It wasn't only the large entry-priced multiples that closed. Several niche retailers ceased trading. Ravel was one of the go-to fashion footwear stores of the 1960s and 70s. It was owned by the Wise family who were shrewd operators, taking stores in prominent corner locations. Ravel was particularly strong at offering the key looks of the season, whether it was heavy platforms or high stilettos. In 1974 it was bought by Clarks. For a few years Ravel continued to lead the way in fashion. Gradually it lost that special something that gave it an edge. Being part of a large organisation, and the need to comply with its corporate processes, started to inhibit innovation and risk taking. The business declined and was finally closed in 2007.

Faith Shoes was a well-run family company led by Jonathan Faith, an understated but inspirational leader. In 2004 he sold the business to the private equity company Bridgepoint, for £65m. After the sale a lot of the key management, including Jonathan, left the company. There was no one left who truly understood the brand. New management was brought in to run the business, but they lacked the same insight and understanding of the Faith customer. The company closed in 2010. Losing key management of a fashion company which leads to their eventual demise has been a common theme in fashion retailing. Ted Baker is a good recent example of how important it is to retain those people in a company who really get the brand.

Shelly's (famous for selling Doc Martens), Derber, Sacha, Read or Dead, all iconic footwear retailers, disappeared from the high street. In some cases, the brand was bought and continued as a wholesale business. In most, the brand just fizzled away. What are the lessons of the demise of so many footwear retailers? Firstly, footwear retailing is difficult and if you are sensible, you avoid it. Secondly, unless you have a real point of difference and offer a product that is special, good value (not low price but good value) and can't easily be replicated, you will struggle. Thirdly, you need a talented team of people who really get the brand, understand the customer and can deliver consistently good customer service in a highly competitive and dynamic market. I remember speaking to one of the Church family after it was sold to Prada. When the Prada team visited Church's offices in Northampton for the first time all the financial and commercial information was ready for them to review. Prada said that could wait. They wanted to discuss two things: product and marketing. To be successful in fashion those two things must be spot on. If they are right, good financial performance will follow.

It was clear that Comme il faut was not an acceptable name. It was too long and customers didn't appreciate a French phrase that wasn't in everyday use and was difficult to pronounce. In the US the founder's name was often used to define the brand. Ralph Lauren and Tommy Hilfiger are household names. In footwear Steve Madden and Sam Edelman are strong brands using the founder's name. In the UK this isn't so popular. Surnames or first names are sometimes used. The popular Swedish brand Hennes & Mauritz has been abbreviated to H&M. The UK footwear brand Russell & Bromley uses the names of the founders, although these names can sound like a firm of lawyers rather than a fashion brand. I wasn't keen on using my name for the brand. One of the important features of the name is that it must be memorable, which usually means a short name. It was with these thoughts that I considered the future of my brand.

16. Dune is born

We traded with Comme il faut for a year. It was time to find a new brand name that was short, descriptive and English. Registering a name is always a challenge as most names have already been taken. The company I had set up to operate our retail business was Warwick Limited (we lived in Browning Close round the corner from Warwick Avenue in West London) but this wasn't an appropriate name for the brand. In the end I decided on the name Dune. I liked the metaphor of the shifting sands of Dune. Just as a sand dune blows in the wind and constantly changes its shape, so fashion is constantly changing and reinventing itself. We were a fashion brand, so it felt totally appropriate. As well as having a compelling story, importantly the name was available, so I quickly snapped it up. So, in 1992 Dune was born.

I had the name and the product, but I didn't have any channels to distribute the Dune range. Opening stores was an expensive proposition and I didn't feel that we had sufficiently proved the concept or had the necessary experience to commit to opening stores. At the time there was a weekly trade magazine called *Shoe*

& Leather News. Reflecting the fragmentation of the footwear industry it has since been absorbed into *Drapers*, the clothing weekly. In the small ads at the back of the magazine there was notice asking for a women's footwear brand to take concessions in Jane Norman stores. Jane Norman was a successful women's young fashion chain with stores mainly in London and the south-east. I answered the ad and arranged for the owner and CEO of Jane Norman to visit our showroom at 9 Hatton Street so we could present the range to him and his team and try to persuade him that we were the right choice for his footwear concession.

Norman Freed, the owner, who ran the business with his two daughters and finance director Saj Shah, was a true gentleman. If you had to choose someone to go into business with then Norman would be at the top of the list. Apart from being a really nice guy, he was direct and decisive. We gave our presentation which went well. The range looked strong and would fit well into Jane Norman stores. Norman was complimentary but, understandably, had reservations about our lack of retail experience. I explained that we had successfully operated a few concessions in footwear stores. I also reassured him that we wouldn't take the concessions unless we were confident that we could manage them effectively and make them a success. After some deliberation he agreed to launch Dune concessions in three of his Central London stores; two on Oxford Street and one on Brompton Road in Knightsbridge, just along the road from the famous department store, Harrods. These were prime shopping destinations. We were given a good position in each of the stores, so there could be no excuses if it didn't work.

In many ways, the timing wasn't great. In 1992 unemployment had reached 10 per cent and the economy had entered recession. But Central London was a bubble and attracted a lot of tourists, both from the UK and overseas. There was no congestion charge or prohibitive parking fees in those days to deter shoppers as there are today (Mayor of London take note). At the weekend the streets in the West End of London were packed with cars coming to the capital bringing people who were keen to spend their time and their money shopping. There was no online shopping to distract them. This was the heyday of physical shopping.

The concessions were a great success. Our shoes complemented Jane Norman's clothing range, which was fashionable, contemporary and particularly strong on dressier styling, where we excelled. Based on this success we were offered concessions in other locations where there was enough space on the shop floor but importantly in the stockroom as well. One of the challenges of footwear retailing is that you need the space to store the shoes. Unlike dresses, which neatly hang on rails on the shop floor, only a half pair of one size is on the shop floor, the rest of the stock is in the stockroom. This was a problem in the smaller Jane Norman stores as there wasn't enough space, even though we tried to overcome this challenge by making daily deliveries to replenish the stock that had been sold. Typically, a third of a footwear store is taken up with the stockroom. You are paying rent on space that is "back of house", that is not selling space. I found out that the value of our stock in our concession in its prime Oxford Circus store was more than all the stock Jane Norman held in that store. They had the advantage of fast deliveries from their suppliers in France and a limited size range. Both meant that they could turn their stock much faster than us and hold a lot less stock.

Going into Jane Norman stores in Watford, Bromley, Croydon and Merryhill Shopping Centre near Birmingham, gave us insights into trade outside London and the need to tailor the range for different locations. The bestsellers were always the best sellers everywhere. However locally we found that some stores were better at selling more casual products whereas others were particularly strong on occasion wear. Understanding these regional differences helped us send a more appropriate range to each of the stores. Not all the stores worked. We went into the Jane Norman store in Romford, in East London. We quickly found that the Romford customer wanted lower prices than we could offer so we quickly exited the store.

Although I was learning about retail, my experience had been in manufacturing, importing and wholesale. No one in the company was a retail expert. We weren't using any science to decide how many pairs to buy for the concessions, it was done purely on "gut feel". Although I am a great believer in gut feel I recognised that we needed to be more analytical in making decisions. We needed someone who understood retail and could educate us in best practice. I was lucky to know Dennis Fleischer. Dennis was a good friend of the family who was very close to my former partner and uncle, Len Goodman, and his wife, Eve. He had worked for many years as a director of the BSC and had recently left Stylo Barratts where he was managing director. He had helped us with the range planning and merchandising of Comme il faut in Barratts and Faith.

Dennis was a tall, slightly hunched man with thinning ginger hair. Apart from being an incredibly generous and warm-hearted individual he was brilliant with numbers and had invaluable retail experience. Not only was he excellent at preparing our sales and stock plans, but he was also a talented teacher with an immense amount of patience. He was very organised and clear thinking. In those days we didn't have a computerised system, we relied on Excel spreadsheets and handwritten schedules. Dennis, who, unlike me, had very neat handwriting, prepared a clear seasonal plan which covered our budgeted sales, broken down by full price and markdown sales, stock requirement, number of options, and the pairs we should buy for each option. For many years we used this template which combined with our feel for what were the strong styles, enabled us to trade successfully and avoid any big mistakes. What I particularly liked about Dennis's approach was that he avoided overcomplicating things. He had that unique skill of simplifying quite complex issues so that we could focus on the priorities, in particular the product, and not get lost in the detail.

Dennis recruited our third employee, a young man called Mohamed Yacoobali. He was working in Barratts' flagship store on Oxford Street. He was persuaded to join us to run the Jane Norman concessions. Mohamed was an energetic, hard-working and driven individual. One of his jobs was to go round all the stores to collect the handwritten sales sheets which Dennis summarised and produced a weekly sales report. Mohamed transferred to our small head office team where he was Dennis's assistant. He gradually took over many of Dennis's responsibilities. He was ambitious and in 1999 he left to work in Saudi Arabia, where he was attracted by the low tax regime, the culture, and the opportunity of earning a much higher remuneration package. In 1993, based on the success of the concessions, we made the decision to open a store. Finding the right location is of prime importance. I was aware of the mantra of retail agents. When considering acquiring a property the three key factors to consider are: "location, location and location".

17. First stores

It was essential that we chose the right location. We wanted somewhere in London with good footfall but with a reasonable rent, which ruled out Oxford Street. In the end we decided on King's Road in Chelsea. It had been one of the key destinations of the fashion crowd in the 1960s and 70s. It was the home of Mary Quant, Biba and Viviene Westwood, the key designers of the period. Although it didn't still have that iconic status in the 1990s, it was a busy street for fashion retailers with a good mix of affluent local customers and tourists. It was also important that we were adjacent to complementary clothing brands like Karen Millen, Hobbs and Reiss.

Dennis and I walked up and down the street many times and decided on the shopping parades near Sloane Square as the further west you went down King's Road it became more niche and alternative, like the shop selling cowboy boots with the memorable name, R Soles. There were two stores available. One larger one on the stronger north side of the street and a small store, 37A King's Road, on the south side. The rent on the larger store was significantly higher but the location was stronger. The other challenge of the smaller store was that the shop front was narrow. There was a danger that customers would walk past the shop without noticing. On the plus side the adjacent stores, apart from the bank next door, were strong brands and appealed to a similar customer to Dune. It was only later that I learnt that taking a shop next to a bank wasn't a good idea as shoppers walk quickly past a bank, as there is nothing to see in the window, and there is a danger that their momentum takes them past your shop as well.

Being risk-averse, and as this was our first store, we went for the smaller one. It gave us the opportunity to test the concept in a modest sized store, in a good location where the occupancy costs were reasonable. At the time we were only selling women's footwear, so we didn't need too much space. In addition, the landlord was reasonable and didn't demand a rent deposit or personal guarantee which many of the larger landlords insisted on, although as we grew most of the landlords wanted a guarantee from Browning as there were no assets in the retail company.

I liked the concept of affordable luxury. The design and quality of a pair of shoes you would buy from a luxury brand but at an affordable price. Our shoes weren't made in an expensive artisan factory in Italy but the quality from José Molina was not so different and you could buy four or five pairs of Dune shoes for one pair of Prada or Gucci, which incidentally is still the case, although now it is more like six or eight pairs as the luxury brands have increased their prices to such a high level. Our designs were fashionable but accessible. We weren't fashion forward. We were not aiming to push the boundaries of fashion, but we were interpreting it in a distinctive Dune way that was stylish, elegant and wearable. Our typical customer was around 30 years old, liked fashion, wanted to look good, wanted quality leather shoes but didn't want to spend a fortune on a luxury brand.

Given the decline of the specialist footwear retailers, opening another footwear store was questionable. There were two reasons why I felt that it was a sensible move. Firstly, we had a clear niche in the market. We weren't competing on price, as this would have been a losing battle. We were going to offer a well-designed, quality product at a fair and affordable price. Unlike many of the specialist retailers that were struggling we had a focused range that appealed to a specific customer group. Secondly, we needed to create a brand. Our experience with Browning had demonstrated the limitations of selling an unbranded range. In 1993 opening shops was the best way of establishing the brand.

The big lesson of selling to the multiple specialists was that if the range you were offering wasn't different then the chances were you could buy that product a lot cheaper in a clothing or department store. I have always felt that the product is the most important ingredient in building a retail brand. If you get the product wrong, even if you have the nicest store and the best sales team, you will struggle to survive.

The core of our range was the shoes from José Molina, but we needed other categories of footwear, like casual sandals, boots and casuals, that they did not make. I went to the Milan shoe fair to find factories that made these products. The main challenge was finding factories that would make the small quantities that we needed for the concessions and our King's Road store, which was around 40 to 60 pairs of a colour of a style. The problem with placing small orders was usually the factory charged a premium price which either made the shoes too expensive or reduced our margin to an unacceptable level. At Browning I was used to receiving large orders from our customers. We tended not to deal with customers who bought small quantities as they were not worth the effort. It took just as long to manage a small order as a large one as the administrative work was the same.

I was aware that many suppliers knew that I owned the import business Browning, which was still going, and were worried that I would take their styles and copy them. I tried to reassure them that was not the case, that Browning and Dune were run completely separately. It was also important that the shoes I bought from these factories complemented the shoes from José Molina so that the range looked cohesive and well planned. I went to the fair with a slightly defensive and nervous frame of mind. Going to ask for small quantities for a newish retail venture that could be a competitor wasn't going to be easy. As it was, I was lucky to find three good Italian suppliers who accepted our modest orders, two of which, Palex based in Tuscany, and Stilmoda near Brescia in Lombardi, we still buy from some 30 years later.

A big decision was the shop design and fit out. We got a couple of quotes from specialist shopfitters that came to around £250,000. This was way outside our budget. I had been travelling to Paris frequently and I noticed that several of the footwear retailers had the shoes displayed on top of boxes. The concept had two advantages. Firstly, you were using the space more efficiently as stock was being held on the shop floor, which meant less space was required "back of house". Secondly, you could serve customers faster. You could take the correct size from the stack of boxes and not have to disappear for several minutes to find the shoes in the stockroom. Being an impatient person, I always get frustrated being asked to take a seat while the sales associate goes to hunt for the correct size. I read somewhere that in studies it was found that after two minutes customers become restless and lose interest. It is certainly one of the challenges of selling shoes, although luckily most customers are more patient than me. I therefore liked a system that meant the customer was served quickly. Thirdly, it was cheap, a lot less than £250,000. The disadvantage was that it didn't fit the concept of affordable luxury. Having stacks of shoe boxes wasn't exactly premium although we did make the effort to make it look attractive. The system was used in France by brands that were in the middle segment of the market and customers didn't necessarily associate it with cheap shoes. I remember when we moved to a more conventional method of displaying shoes, we had several complaints from customers who said we had lost the accessible, inviting character of the on-top-of-boxes concept.

I decided on an ancient Roman theme for the store. An unusual choice and maybe not my most inspired creative decision. I had seen these impressive wall fixtures of Roman columns and characters in a terracotta colour in a Paris shop and thought that they would work well on the walls of the stores. I would place plants on them to add some colour and warmth to the store. I painted the walls in a mottled burnt orange colour which complemented the Roman theme. The fixtures and stools were in a dark anthracite steel with frosted glass on the fixtures. The most expensive item was the limestone floor. The shoes were in white boxes with the black Dune logo and were stacked around the sides of the shop and under the fixtures. The overall effect was minimalist, bordering on brutalist. However, importantly, the shoes were the hero, and the concept made it easy for the customer to shop. The fact that there was nothing like it on the high street made it stand out, even if sometimes it was for the wrong reasons. It was exciting that we now had our first store and seeing customers come into the store, try on the shoes and buy them. I still get a thrill when I see someone carrying a Dune bag and get even more excited when I see them wearing our shoes.

The initial sales were disappointing. We were taking more in the Jane Norman concessions than in King's Road. The feedback from customers on the product was positive and we didn't get any adverse comments on the shop fit although some older customers complained that the stools were hard and uncomfortable. The reality is that it takes time for customers to embrace a new store. We hadn't done any marketing, so we were reliant on passing trade. After six months or so we started to see footfall and sales increase. It wasn't making a fortune, but it was sufficiently good to encourage us to look for another in a similar location. Unless you are in the shop serving, doing the accounts and generally devoting your energies to the shop, which is what a typical independent retailer does, often with support from the family, it is difficult to make a profit from one store. You need a chain of at least six shops to get the necessary economies of scale. Our next shop was on Kensington High Street, a busy shopping street in West London, which we opened in 1994. Unlike King's Road it had a wide frontage but was very shallow. There was an old-fashioned wrought iron circular staircase to go downstairs to the stockroom which made going up and down stairs tricky. Only one person could be on the stairs at any one time. It made serving customers a challenge and only the younger, more athletic of our sales team were comfortable working there. It was one of the reasons that the rent was reasonable. It was a highly visible shop. As you exited Kensington High Street underground station and looked right you could see our store which I thought was impressive.

Our third store was in a large shopping centre in Thurrock in Essex, east of London, called Lakeside, which we opened in 1996. This was a larger store in a prime location in a major shopping destination. The customer was different from King's Road and Kensington High Street which typically attracted more affluent Londoners. Lakeside has a more diverse customer base and has a much larger catchment area extending into the county of Essex. "Essex Girl" was an unfortunate stereotype that started in the 1980s and 1990s which *Time* magazine described as "a lady from London's eastern suburbs who dresses in white strappy sandals and suntan oil, streaks her hair blond, has command of Spanish that runs only to the word "Ibiza" and perfects the air of tarty prettiness". That did not typify our customers in Lakeside, although we did sell a lot of dressier and going out styles and quite a few white strappy sandals.

In those days it was popular to have a celebrity cut the ribbon and officially open the shop. Appropriately we chose Patsy

Palmer, one of the stars from the popular soap, EastEnders. There was a large queue to get into the store to meet Patsy. She was the perfect celebrity, friendly and very much at ease talking to customers and signing autographs. We stayed in the same shop for 28 years when we relocated to a larger shop in the centre. Lakeside was an important test. It gave us an opportunity to see if the Dune formula worked outside Central London in a very different shopping environment, and fortunately it did. Although we traded well in Lakeside when I compared our sales with Jane Norman's, which had an identical store opposite ours, it was slightly depressing. Their sales were much higher than ours. It demonstrated how a successful clothing store, with a wider range of products, could achieve higher densities than a shoe shop selling one brand.

The success of Lakeside gave us the confidence to open in other shopping centres and cities outside London. The main criterion for the stores we chose was the location. We wanted to be adjacent to similar brands such as Karen Millen, Reiss, Ted Baker and Whistles. The shop size was to be around 1,600 square feet. We needed a location that had a good flow of customers. Finally the occupancy costs had to be reasonable. I made a few mistakes that were costly. In Meadowhall, a large shopping centre outside Sheffield, we took a shop off the main concourse and paid too much rent. We had a ten-year lease and the rent, which was reviewed every five years, was upwards only, which meant that even if the market rent went down, which it did, we were still paying the higher rent. The result was that we were stuck with an unprofitable shop for ten years. Fortunately, Meadowhall was an exception and most stores, in cities like Glasgow, traded well and made a good profit.

18. Browning Struggles

While setting up and running Dune, I was still focusing on Browning. We needed Browning to be profitable to fund the growth of Dune although it was becoming harder all the time. As the factories started to deal directly with UK retailers our margins as an importer came under more pressure than ever before. I decided we should try to source some of the shoes straight from the factories in an effort to increase our margin. We were working well with Janna. Maxgreat continued to be an important supplier. However we saw an opportunity of sourcing a cheaper product from factories in the north and east of China which produced a more competitive product.

In 1999 we set up Browning Hong Kong, our first overseas subsidiary. The Hong Kong company name was Gwek Ling, which I was told by our agent meant "dancing feet", which was appropriate. I am not sure where he got that translation from, but I understand it means "clever" or "nimble". The main function of the company was originally to deal with shipping issues. Some of our customers wanted to ship FOB (Free On Board) from Hong Kong. This suited us as it meant that we didn't have to ship the goods to the UK and incur the costs of handling and often storing the goods until they were required by the customer. We just handled the consignment to the customer's agent in Hong Kong, and they arranged shipment. It also meant that we got paid faster which was good for cash flow. There was also a cost advantage of arranging the shipment from Hong Kong rather than doing it from the UK. We set up a small logistics team in Hong Kong which was less expensive than having the team in London, and they were often able to negotiate better shipping rates locally.

The less we had to handle the goods the better. If the factory could load a container at their premises and deliver it directly into the customer's warehouse, that was ideal. If the factory had to deliver into our warehouse, we had to sort the orders, arrange them for loading onto the container before delivering to the shipping company. This incurred additional costs and often adversely affected our margin. It was often difficult to persuade the customer to buy FOB as they liked the convenience of us handling the shipping. They often delayed taking in the stock to help manage their cash flow. Not only were we designing the product for our customers, sourcing it from an excellent factory, checking the quality, shipping the goods but, for some customers, we were also helping them manage their cash flow. All this for a measly ten per cent margin that was coming under increasing pressure.

We started to use Browning Hong Kong to source products from China that Maxgreat didn't offer, such as cheaper women's sandals, casuals and men's shoes. When I started Gwek Ling I appointed a charming lady from Hong Kong, called Lily, to run the small operation. Unfortunately, her knowledge of factories that suited our requirements proved limited. She introduced me to the Chinese Animal By-Product Commission, a body set up by the Chinese government to encourage trade from China. Footwear evidently came under their broad umbrella of animal byproducts. It didn't sound like a very encouraging introduction, but I was happy to explore it further. We were flown to Shanghai to visit some state-run factories making footwear that they thought might be of interest. These were huge factories with thousands of employees, all dressed in identical yellow uniforms, making army boots. It wasn't what we were looking for.

The highlight of the trip was dinner at a fish restaurant in Shanghai which was attended by their delegation of ten people, including the chairman of the department (who looked like Mao Zedong, the Chinese revolutionary leader), Lily and myself. After several toasts there followed a series of beautifully prepared dishes. When the pièce de résistance, a massive fish on a silver platter, arrived, the head was quickly cut off and placed in front of me. Being the guest of honour it was my treat. I eat most food, but I wasn't attracted to this large head with its eyes staring at me with its mouth open revealing a set of razor-sharp teeth. I thanked my host but insisted that he have the fish head as he was the most senior person present. He quickly accepted this offer and proceeded to devour the head, eyes and all.

Lily wasn't the right person to run the company, and we soon parted ways. We then worked with an agent called Eric Tse. Eric had been in the trade for some time and came with excellent contacts. He had worked for ISA in the UK so knew the market well. We had some success with Eric until he decided to go it alone, eventually teaming up with David Brook, our former sales director.

We continued to look for factories to add new products to the range. Jason and our quality control team were tasked with this job. They were often joined by our UK team on their trips to China. These products added a new dimension to our business and enabled us to grow, and service the supermarkets and discounters which were becoming a serious force in selling footwear. They were used to working on tight margins, placed big volumes and could therefore sell the products at competitive prices. The downside for us was that this was a cutthroat business where margins were tight. The factories were not of the same standard as the factories used by Maxgreat so there was an added risk of quality issues. Most of our quality controllers' time was spent inspecting orders at these factories as a quality issue was particularly costly given the large volumes and low margins. The UK retailer placing the biggest orders was Primark. They were shrewd as they placed their large orders early and used this to negotiate very low prices. In many cases the factories were taking the orders at minimal profit but were happy to use their orders to fill the factory and cover their overheads.

It wasn't a business that I enjoyed. The factories were of an acceptable standard, large sheds in the middle of nowhere, but seeing these budget-priced synthetic or fabric shoes going round the production line didn't fill me with pride. I questioned whether this was the sort of product we should be involved in. Unfortunately, the market generally was getting more and more competitive. The specialist footwear retailers were looking for lower prices to try and compete with the non-specialists whose cost base and prices were a lot lower. In the end, as we were to discover, they couldn't compete and most of them disappeared from the high street.

During 2005, as our margins came under pressure, we made the decision to replace some of the products we were buying from Maxgreat, and other local trading companies, with goods we sourced ourselves. The argument being that the ten per cent (or more) that Maxgreat was charging for their services we could do ourselves at a lower price. We employed a sourcing director, David Hampson, who was based in Hong Kong. He was tasked with building a team, including setting up a sample making plant in China, so that we had our own vertically integrated operation. David was an experienced shoe man. He had worked for many US footwear brands in Hong Kong and most recently for a large Chinese manufacturing company. He was a physically imposing man, well over six feet with a shiny bald head. He was good company with a friendly demeanour and a confident personality.

We took offices in a Harbour City overlooking the harbour where we had a showroom for visiting customers. David built a team that grew to 20 people. We relocated some of our UK departments to Hong Kong as, due to the lower wage rates, it was cheaper to operate them from there than London. They joined our logistics, sourcing and quality teams. It wasn't a success. We underestimated the added value that Maxgreat gave to our business.

Building a strong team from scratch was a bigger challenge than we anticipated. We had worked with Janna over ten years and both she and her team understood our processes and, more importantly, our customers. She operated an efficient and impressive operation that we were naïve to believe we could easily replicate in a year or two. The samples, the key ingredient to successful sales, that came from Maxgreat were spot on. The samples we got from our Chinese sample plant were often disappointing and needed frequent re-working. Unsurprisingly it was taking time to train the operatives to get to the right look and feel of the shoes.

We also underestimated the challenges of working with and managing a new team that was 6,000 miles away and seven hours ahead in time. Hong Kong didn't get the support needed from the management team in the UK to train and integrate the two operations into one unit. My conclusion was that it is hard enough to build and manage a local team in London. To do it successfully in both London and Hong Kong, so that the offices worked seamlessly together, required a clear plan and excellent execution, which meant frequent visits by management to Hong Kong, which wasn't happening. It also would take time, and time wasn't on our side. We continued to operate the Hong Kong office until 2008 when we gradually closed it. It was a good lesson in the difficulties of operating a remote office. A lesson that I didn't learn well enough when we opened our Dune office in New York in 2016.

As I got more involved in building Dune I was travelling less often to Hong Kong and the Far East. I visited every six months to meet Janna, visit our offices and the factories and spend time in Hong Kong or Tokyo going round the stores looking for inspiration. I flew into Hong Kong in December 2013 to be met by Jason. We were to drive to Dongguan to meet Janna. I joined the queue at passport control and showed my passport. The officer spent some time studying the passport. He eyed me suspiciously and asked me to wait. Eventually two policemen arrived and asked me to follow them to a waiting room where I was told to take a seat. In the room were a group of African women in bright tribal dress, a couple of women with very short skirts, bright red lips and a surfeit of make-up, a man with a large angry looking scar on his cheek and me. This was the room for those with problems.

After half an hour a police officer came and escorted me into a small room and asked me to confirm my name and the purpose of my visit. I asked him why I was being detained. He ignored my question and asked me to wait. Another policeman came into the room, who looked more senior as he had more stripes on his jacket. He announced that I was under arrest. I didn't have to say anything but if I did it would be written down and used as evidence. A lot of thoughts went through my head, the most pressing was why was I being arrested. Was this a case of mistaken identity? I couldn't think of any offence that I had committed in Hong Kong.

The policemen told me to follow them. We went to the customs hall where the officer proceeded to carry out a detailed examination of my luggage. Every piece of clothing was taken out of the case and inspected in minute detail. I supposed this was standard treatment for someone who was under arrest. After customs I was ushered into a police van and driven to the airport police station. As I was going into the station there were two hooded men in hand cuffs being escorted out – not a very reassuring sight. All my possessions, including my mobile phone, were taken from me, and I was placed in a bare cell, where previous inmates had scratched their names and messages on the wall. I was starting to feel uncomfortable. The offence must have been serious for me to be treated in this way.

I had been able to make two phone calls before my phone was taken away. I had alerted Jason, who was waiting in arrivals, of my predicament. I had also called a friend in Hong Kong to ask him to arrange for a lawyer to represent me and hopefully get my release. While I was waiting in the cell, I was racking my brain to try and understand why I had been arrested. I couldn't think of any reason unless there had been some impropriety at Gwek Ling that I had not been alerted to. I am not a nervous person, but I was starting to feel a mixture of concern, frustration and anger. You hear of innocent people languishing in prison for days for a crime they didn't commit.

After an hour in the cell, I was taken into an interrogation room. Up to this point none of the policemen had spoken English. A translator had been brought in which was helpful. In answer to the question as to why I was being held the answer was "money laundering". This wasn't a great help. When they explained that the amount was HK$231,534 (about £20,000) I felt a sense of relief. At least the amount was modest. This wasn't a major case of money laundering. I still wanted to understand why I was being held and when I would be released.

Two plain clothes policemen appeared and told me to accompany them. I was being taken to the main Central station on Hong Kong Island. I was given my phone back and managed to call Jason to tell him where I was going. I also called the lawyer to ask him to come to the station. When we arrived at the station I was put in another cell for two hours. They were not in a hurry to deal with my case or explain why I had been arrested. In the car I had appealed to the policemen to let me have more information as to why I had been arrested. They were plain clothes policemen. They didn't know anything about the case.

165

Eventually I was placed in an interrogation room where a young detective, who spoke English, asked me to explain why I was in Hong Kong and what was the nature of my business. I told him I had been coming to Hong Kong for many years to source footwear, that I was a businessman and was en route to see my supplier in China. I showed him some of the Dune brochures which he studied carefully. In the meantime, my lawyer had arrived. He introduced himself as Philip Swainston. He was a large young Englishman who looked like he had been in a fight. He had recent scars on his face and head. He explained that he had fallen down the stairs of the double-decker bus on the way to the police station. To an outsider he seemed the more likely of the two of us to have been arrested. He didn't appear as if he had handled this type of situation before. I would guess he was sitting in the office and had been selected to represent me based on his availability rather than his experience. His advice was not to say anything which seemed to me exactly the wrong way to deal with the situation.

Another policeman came into the room who introduced himself as Detective Constable Mok. After being questioned by six different policemen, all asking the same questions, I had finally met the detective who was handling the case. He took down a formal statement which he meticulously wrote down and asked me to sign multiple times. Although it wasn't explained to me until later when I managed to get in touch with David Hampson, it appeared that the receptionist at Gwek Ling, a woman called Agnes, who was also David's PA, had been raising false invoices for travel and had been paying the money from Gwek Ling into her boyfriend's company's bank account. David had reported the crime to the police, and although Agnes had repaid some of the money the case was still outstanding. As I was the last remaining director of Gwek Ling I was, in some way, considered responsible. After two hours of questioning, I explained that I was not involved in this affair. I insisted on being released. Detective Mok said he understood and left the room.

I consider myself a level-headed and reasonable person but having been held for ten hours, with only a bottle of water, and repeatedly being asked the same questions, I was losing my patience. No one seemed to be interested in my predicament. It was approaching 10pm and there seemed to be a general exodus from the building. I was told that I would have to spend the night in a cell until the matter was resolved the next day. I said under no circumstances would I spend a night in their cells. I insisted on seeing Detective Mok. I got the sense that he knew that I wasn't responsible for any crime. He came into the room in his coat, as he was just about to leave for the night. I pleaded with him. What would it take to be able to leave the police station? I had clearly not committed any offence and there were no grounds for holding me. He was sympathetic and finally agreed that if I could produce a bond of HK$100,000 (£10,000) in cash he would release me on bail. The only people in Hong Kong who could meet that request were the Kongs, who lived in Hong Kong, and were a major supplier to Dune. I spoke to John Khuu, their son-in-law, and told him what I needed to get out of jail. He said he would do his best. It was a lot of money to withdraw from an ATM. He arrived at the station an hour later with the cash, after he, and his family, had visited most of the ATMs in the city. I felt a huge sense of relief. I was finally a free person. I met Jason, who had been waiting patiently for me all this time, and we drove to Dongguan, 12 hours later than we had planned.

The next morning, I spoke to David and explained what had happened. He was very apologetic and said he would speak to the police to sort it out, which he did. The following week I returned to the station with John to collect the money. After that experience I was always nervous when I arrived at Hong Kong airport even though I had insisted on a letter from the Hong Kong police confirming that I was not involved in money laundering and was a respectable businessman. Fortunately, subsequent trips were incident free, although I always had a slight worry approaching the customs barrier.

19. Becoming a multiple retailer

By the early 2000s, I still had the two businesses. Browning Enterprises, whose margins were increasingly being whittled away, was the most profitable at the time. But Dune was my dream and the major focus of my efforts. I felt the future was in having a shoe brand, rather than being an importer. So I was ploughing the profits from Browning into Dune to grow the stores

I needed a bigger team to expand Dune. Our original head office team consisted of Dennis, Mohamed and myself. Dennis managed the merchandising and administrative functions and I did the buying. By 2003 we had grown to a team of 30. Sadly, Dennis Fleischer had died of cancer in 1997, in his late sixties. He had been invaluable in guiding us through the early years. I had learnt an enormous amount from him about how to run a retail company. I missed his advice, quiet authority and friendship. He was a meticulous man with an amazing aptitude for numbers. He was a creature of habit, rarely changing his routine. He had a plain cheese sandwich on white bread with a cup of black coffee every lunch during the five years he worked at Dune. He couldn't be tempted with anything more exotic, not even a slice of tomato.

Three of the team that joined us before 2003 stayed with Dune for many years. Barry Marshall was appointed Finance Director. Apart from being an excellent finance director, unlike many finance professionals, he was passionate about the brand and the product. It was great to have someone with such enthusiasm and loyalty in the team. He was certainly a driven individual taking part in several Ironman tournaments which sounded brutal.

Zoe Brookes started as manager of our Meadowhall, Sheffield, shop, was promoted to Area Manager and then became our Retail Director. Zoe was an inspirational leader with an indefatigable energy and drive. She built a team that was incredibly loyal and motivated. Her mission was to give amazing customer service, and she was hugely successful in that endeavour. Customer feedback on Dune's service levels has been consistently high and that was largely due to Zoe's hard work and dedication. It was said of Zoe that if you sliced her in half like a stick of Blackpool rock, the word "Dune" would be embedded through her length.

Jamie Brogden joined initially as our PR and Product Development Manager. He had worked in PR for fashion magazines and joined from Faith where he had successfully held a similar role. Jamie is one of these people who is boundlessly upbeat and enthusiastic, as well as being very talented. His role quickly expanded to become our Head of Creative, a position he still holds. He was responsible for both creating the ideas and managing our seasonal marketing campaigns and heading up our design team. Jamie is the doyen of the dress product. Some of the heels and styles he designed are works of art, not always commercial, but hugely influential in projecting the right image of Dune and elevating the brand. He is one of the people in the company who really gets the brand and the product, which is immensely important. Companies decline when they don't have a team that has a deep understanding of the brand personality as well as instinctively knowing what our customers want.

Having people like Barry, Zoe and Jamie, who shared my vision and passion for Dune, was essential in those early years and gave me the confidence to grow the company. It is a platitude that the people are the most important asset of a company, but it is so true. As the company grew, I realised that if I could bring in talented people to run the different departments my life would be much easier. I also recognised that I was good at certain aspects of the business but not so good at others. I was beginning to learn that as CEO I needed to step back and not get too involved in the detail and leave it to the manager responsible. I often found this difficult. They say that "retail is detail" and it is true that getting the detail right enhances the customer experience. What it shouldn't do is distract you from the key priorities of the business.

Our biggest mistake in the early years was when I took on a new buyer from the young fashion brand, Topshop, which was the hot brand in those days. I allowed her more freedom than usual in building and buying the range. It was a disaster. It was too young and aggressive, and our customers didn't like it at all. It was a big lesson for me of how easy it is to get the range wrong and how costly a mistake could be. After that episode I attended all the range reviews to ensure it was consistent with the brand and right for our customers. Fortunately for the past 12 years we have had an excellent Design and Buying Director in Debra Bloom, who totally gets the customer and the brand which has made my life a lot easier.

In the first few years we learnt what our customers liked about the brand, and particularly the range. We learnt to our cost that she didn't like young aggressive styling. She liked styling that was feminine and easy to wear but with an attention to detail that made the product special. She certainly didn't like "tricky" styles. She wanted to look good, but she was more of a follower than a leader. We divided our female customers into three profiles. "Sarah" was our entry-priced customer. She was a

Next or Clarks customer who aspired to buy Dune and bought our entry-priced product. She was a more considered customer who tended to buy styles that were not over fashionable and which she could wear with several outfits. "Lucy" was our core customer. She wanted comfortable, stylish good quality footwear that was mainly in leather. She felt that Dune offered her this. She bought into different parts of the range, although she was particularly keen on our boots in the winter and sandals in the summer. The challenge with our Lucy customer was to persuade her to buy from Dune several times a year. "Amber" was our fashion forward customer. She was more adventurous than Lucy and less loyal. She knew what she wanted and would only buy from Dune if we had that special product that she loved. Interestingly my daughter, Olivia, who has her own fashion label, is a typical Amber customer. She aspires to buy the "in" luxury brand but buys (or rather takes) from Dune when we have something more trend-led that catches her eye.

With our brand developed, we needed to have more outlets. The problem was that opening shops was so expensive. The combined investment in the shop fit and stock was about £350,000 for each store. Over the years Browning provided more than £10m to fund Dune's growth. Without the support of Browning, we would have struggled to open so many stores. I do sometimes wonder whether spending so much on the shops and continuing to do so for the next ten years, was a good commercial decision. At the time I never gave it a thought. I was ambitious to create a successful chain of stores.

One idea was exploiting the success of international brands. In 2002, I was approached by Aldo Bensadoun, the owner of the footwear chain Aldo, to launch the brand in the UK. Aldo Bensadoun established his footwear chain in Canada in 1972. He came from a Jewish footwear family from Morocco. His father was a footwear retailer in Morocco and later moved to France. Aldo emigrated from France to Canada as a young man and after working in retail for a few years he set up on his own. Based in Montreal, Aldo had been very successful both in Canada but also internationally, especially in the US where there were hundreds of stores. At its peak it had 1,600 stores in 80 countries with a turnover of $1.8bn. I visited Aldo's impressive headquarters to meet the team and look round the stores to get a better understanding of the brand. Aldo was like a cheaper Dune but more mainstream and commercial. Its men's product was particularly strong and contributed a significant percentage of Aldo's sales. I arrived at its one million square foot campus and was struck by the imposing reception area with its glass ceiling and massive olive tree. As they say, you don't have a second chance to make a first impression and the first impression at Aldo was strong. After meeting the team, I was taken around the stores by Mr B himself. It was clear he was an inspirational leader who was held in very high regard by everyone in the company, in particular the retail team, who warmly welcomed him into their stores. Overall, I was impressed with what I saw. This was a slick, successful operation.

I returned to London and discussed the proposal with my colleagues. We decided that Dune was our priority. Despite being an attractive proposition, we didn't need distractions. We knew that international brands also tended to find it hard to expand into the UK's competitive and challenging market. Gap, J Crew, C&A, Forever 21, Toys R Us are just a few that have failed to crack the UK market. In fashion footwear, apart from the sports brands, I can't think of many footwear brands that have successfully launched in the UK with the exception of single-product brands like Crocs, Birkenstock and Ugg. In general, I was guilty of succumbing to distractions but understood that we needed to devote all our energies to building Dune, not launching Aldo. And we did. By 2003 we had 30 shops covering most of the UK. We had become a multiple retailer. Most of our stores worked well. Some didn't. For some reason we never traded well in Newcastle (over the years we tried three different stores in the city and none of them worked). Shops in smaller towns like Watford, Bromley and Bath were also disappointing. There simply weren't enough customers coming into those shops to generate the sales we needed to make an acceptable profit. The other major disadvantage was that these smaller turnover shops held too much stock, which drained cash. Because the shop had to have a reasonable number of styles to make the offer attractive to the customer, and each style had a minimum of seven sizes, these smaller stores turned their stock too slowly.

We traded well in major shopping locations with high footfall and the right adjacencies – the neighbouring shops. Although the occupancy costs of these sites were high, we could still make a good profit. One of the biggest elements of the occupancy costs was (and still is) the business rates, which were usually around half the rent. Only in the UK are retailers lumbered with this iniquitous tax, which makes opening a store an unattractive proposition, especially for a small independent retailer, and leads to some town centres having a lot of charity and coffee shops as well as many that are closed and boarded up.

Having stores in prime locations was also an important brand statement and acted as beacons for Dune. When the online business started to grow, we found that specific shops boosted online sales in their area by up to 20 per cent. Zara is a good example of a hugely successful company that does not spend a lot of money on brand marketing because it has large, impressive shops in the key locations which act as their primary brand statement.

Eventually the time came when the on-top-of-box concept was not acceptable for the brand. When we pitched to go into Bluewater, the major new shopping centre in Kent, the landlord made it clear that this concept wasn't elevated enough for its prestigious new centre. I realised that designing the new shop was beyond my creative talent, so we employed a Scottish company, Skakel & Skakel, to work with us on a new design for our Cambridge shop which we opened at the end of 1997. They liked the image of the sand dune and wanted to bring that theme into the shop fit. The shop was painted a yellow sand colour. We had a curved ceiling raft with blue halo lighting that reflected the undulating shape of a sand dune and the blue of the sky. To complement the ceiling we had curved floor and wall fixtures, green frosted glass shelves and a light limestone floor. It was a big improvement aesthetically on the Roman minimalist concept that I had created. We rolled out this concept for our new stores including Bluewater which opened in 1999.

In the opening years of the 21st century the economy was showing steady, if unspectacular growth. Tony Blair had won an unprecedented third term as prime minister for the Labour government in 2001. Despite the various challenges Dune had established a niche in the competitive footwear market. We felt with a strong product range, a good understanding of our customers, and with a stable economy, we were well placed to continue to grow the business.

In 2002 our achievements were recognised by the industry. We were voted Multiple Footwear Retailer of the Year by *Drapers*, the magazine for the fashion industry.

An advertising campaign for Dune in 2012, shot in South Africa

Left: A visit to Indian stores, 2019

Above: 'Headline' woven ballerina shoe

Right: 'Nature' shoe and 'Blooms' bag

Below: Stratford 2025 – a new store concept elevating the brand

Above: Collaboration with my daughter Olivia, 2011

Left: Autumn/winter 2025 advertising campaign featuring style 'Matias'

Below: 'Salisburry' shoe, from the premium collection

Above: With my mother Dorie and wife Anne after receiving Lifetime Achievement Award in 2014

Right: Interview with the industry magazine Drapers

FOR ALL THE FASHION BUSINESS

Drapers

MAY 3 2014 £4.75   WWW.DRAPERSONLINE.COM

FOOTWEAR & ACCESSORIES AWARDS
THIS YEAR'S WINNERS & FINALISTS
See centre pages

SHOES MAKETH THE MAN

Dune founder Daniel Rubin reflects on a lifetime of achievement in the footwear industry

DEPARTMENT STORES John Lewis gears up for another 150 years **P24**

ECOMMERCE German website Mytheresa is a go-to for luxury **P27**

ETHICAL FASHION The feel-good brands that look great too **P28**

Above: Campaign with Gillian Anderson, face of the brand in 2020

Left: Receiving a doctorate from the University of Kent, 2022

An autumn/winter advertising campaign featuring the 'Selinni' boot and 'Deliberate XL' bag

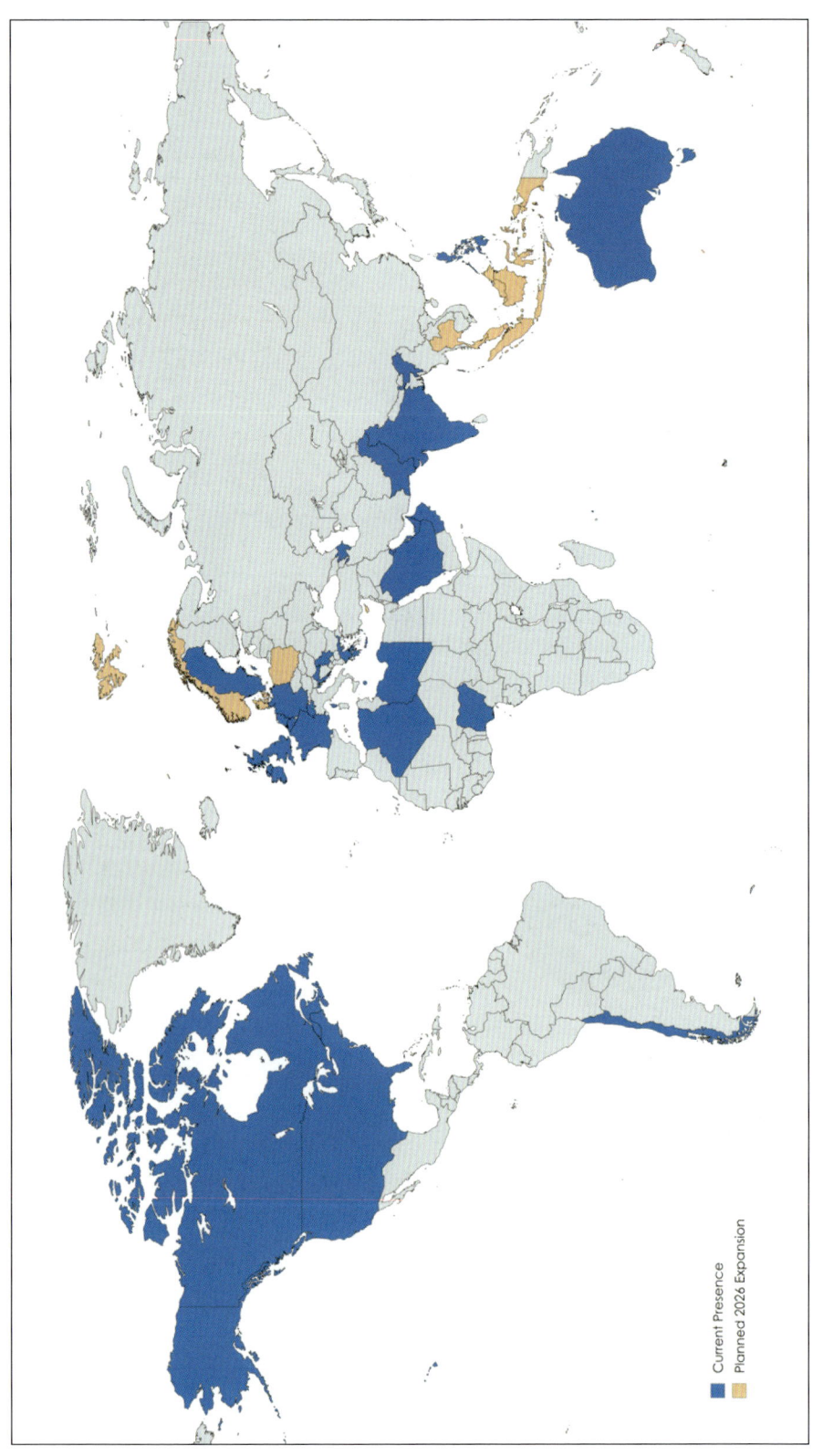

Dune has more than 200 stores around the world. More are planned (in gold)

Current Presence
Planned 2026 Expansion

20. Dune grows

Our stores were making a reasonable but unexciting profit. I realised that having women's-only footwear stores was not going to generate the growth that we needed to justify the investment we had made. It was too niche a business. I was also ambitious. I wanted the company to expand at a faster rate. We were getting to the stage where it was difficult to find new locations for stores that would generate enough sales to be profitable. As we had discovered to our cost, stores in smaller locations didn't work. We had to look for other ways of increasing our sales and improving our margin.

In 1999 we launched a small Dune men's range. If we could attract a male customer into our stores (or a woman buying for her partner) that would increase our sales and profits without materially increasing costs. When I looked at who was selling men's footwear at our price and quality, unlike women's, there were not so many competitors. The market for men's shoes was more polarised with retailers selling at the cheaper or expensive ends of the market. Not many were selling good quality fashionable shoes at around £100. We employed a specialist men's buyer and started to build a range. I had learnt from our previous experience that we needed a range that was true to the brand and sat comfortably next to our women's range. Buying random brands hadn't worked. We needed a clear Dune aesthetic.

Initially we sourced the range from Portugal and Italy. Portugal had good quality men's shoes, although the emphasis was on casual styles. We bought the more formal shoes from a factory in the north of Italy, near Verona. Because we were not reaching minimum quantities, we had to buy from the supplier's range. We persuaded the factory to change some details so that we had an element of exclusivity. As we became more confident, and the quantity increased, we began to develop our own lasts and patterns. Formal shoes became our most popular category and represented 60 per cent of the sales. Today it is 35 per cent, an indication of the casualisation of clothing. Like our female customers, our male customers were not fashion leaders. They wanted a formal brogue or loafer but on a last that was more fashionable than a typical work shoe. Our lasts were less conservative. Some were more pointed, and others had features such as walling round the toe of the last to give it greater character. Although most of the styles were in black or tan, we found a niche for subtle touches of colour such as burgundy or navy, and special materials like patent and reptile prints for occasion wear.

As the Italian factory became too expensive, we moved some of our formal shoes to a factory in China called Artstar. I had found the factory on a visit to the Dusseldorf shoe fair. I went to inspect the factory on my next trip to China. Even getting to the factory was a challenge as it was a five-hour drive from Guangzhou. The building was an empty warehouse in the middle of nowhere. Its operation was like a journey back in time. Every operation was done by hand. The shoes were lasted by hammering nails all round the upper onto the insole. The uppers were then kept on the last for several days to give it a well-defined shape before the nails were removed and the sole attached. The final product was excellent.

We also bought shoes from a Dutch company called Shoes Unlimited which were made in Portugal. Shoes Unlimited designed a range of casuals and hybrid styles (a more formal style with a casual bottom) that sold exceptionally well. A boot called Simon is still one of our best-selling styles.

We started by placing the men's range at the back of the stores as we weren't confident to give it space in the prime area at the front. We were worried that the sales of the store would go down as we would take less money from the men's range than the women's. We soon discovered that unless the men's range was at the front of the store, in a prominent and visible position, men didn't take us seriously as a destination for their footwear. One of the big lessons I have learnt over the years is that you cannot do things half-heartedly. Either you do it as well as you can and persist until you get it right or you don't do it at all. It was certainly the case with men's footwear. Once we offered a well-designed, great quality range of men's shoes in a prominent position in the store, it started to sell well. Men's footwear now represents 30 per cent of our store sales.

Interestingly we made the same mistake when we offered handbags and trainers. It wasn't until we put all our efforts into getting the product right and then displaying the range with confidence that we saw the sales take off. There is the danger if you don't get it right the first time that you give up. Persistence is a great quality. If at first you don't succeed, try, try and try again. There is of course a counter narrative. Fail fast. If something doesn't work, stop doing it quickly because you will save a lot of time and money. Be agile and react quickly to failure. The answer is it depends on what you are talking about. For us, men's shoes needed to be a core part of our range that we had to get right. On the other hand, many years later, we developed a range of children's shoes. Due to space restrictions, we weren't able to sell this range in store, only online. The range looked great, but we soon found out that selling children's shoes online was difficult. Mothers want to try shoes on their kid's feet. After two seasons we discontinued the children's range. This wasn't going to be a core part of the range. It was a distraction. It was right to fail fast in these circumstances.

The other opportunity for growth was concessions in department stores. These were attractive because there was no rent or rates. Commission was payable on sales. During quiet months, like February, when we typically lost a lot of money in our own stores with their fixed costs and low sales, the occupancy costs of the concessions fell in line with the sales, which meant we didn't lose money in these months. The only problem was it was very difficult to get a footwear concession in the department stores. There was a duopoly with two companies, Kurt Geiger and Shoe Studio, controlling the footwear space in the large department stores. Retailers were reluctant to give space to a newcomer like Dune. Kurt Geiger and Shoe Studio also offered other brands to broaden their offer and appeal to a wider customer base. We only offered the Dune range. I had many meetings with Don McCarthy, the charismatic CEO of House of Fraser, and although he was sympathetic, he concluded there was no room for Dune. I had the same response from Rob Templeman, the CEO at Debenhams, another strong personality.

Kurt Geiger, a luxury British footwear and accessories brand, had a special relationship with Harrods and Selfridges, where it not only had its own concessions but also represented many of the luxury brands' footwear collections. The chance of getting into these premium department stores was minimal. We did manage to get a concession in John Lewis in 2005. Following its success, we went into a further ten of its stores.

By 2009 we had 50 stores and had established ourselves as a key player in the footwear market. A lot of our competitors had closed, and we were one of the few independently owned players remaining. Dune was established. But I wanted us to be not just a British brand, but a global one.

21. Going international

Unsurprisingly, establishing a brand in a foreign market is difficult, especially in large, highly competitive markets like the US, Europe and China. The key to success is that elusive quality: point of difference. Most large markets don't need another footwear retailer. There are more than enough companies selling shoes already. The customer is spoilt for choice. The only way you can succeed is if you are offering a range that is different from what is in the market already, and you have a brand proposition that is special and distinctive. At the same time the range needs to be commercial. It is easy to develop a range of leading-edge fashion styles but unless it has broad appeal it is unlikely to sell well. You also need to be aware of local factors – as we discovered in Saudi Arabia.

Saudi was where we started our global expansion, in 2002 with a company called Alhokair. Mohamed Yacoobali, who had worked for us and emigrated to Saudi in 2000, introduced me to the company, where he was working, and the owners, the Alhokair brothers. The driving force behind the business was Fawaz, a charismatic leader who, apart from being a dynamic businessman, had an engaging and charming personality. He worked with his brothers, Salman, an architect, and Abdul Majid, a dermatologist. They had secured an important agreement with Zara to open stores in Saudi and were keen to grow their fashion franchise business. They liked the Dune brand and felt there was a good opportunity in footwear. Saudi women were limited in the clothing they could wear, as most were required to wear the burka, so footwear and bags were an important way of making a fashion statement. Alhokair opened Dune stores in the capital, Riyadh, as well as the cities of Jeddah and Dammam.

I made my first trip to Saudi Arabia in 2004. In those days it wasn't a very welcoming country. Getting through passport control and customs was a harrowing experience as the officers in charge were unfriendly to the point of being aggressive. I hadn't planned my trip very well as I arrived during the holy month of Ramadan which meant that I could only eat my meals in my hotel room during the day, as everyone was fasting between sunrise and sunset. I visited our store in the Al-Faisaliah Centre in Riyadh. While I was in the store there was a commotion. One of the customers in the store had let her hijab fall slightly down her face and a member of the religious police (the Mutaween), who was responsible for enforcing the strict religious customs of Sharia law, had noticed this and was remonstrating with her and asked her to leave the shop and come with him. I was shocked and asked what was going to happen to her. She would only get a warning I was told, which was mildly reassuring.

The stores were run by immigrants from the Indian subcontinent and Southeast Asia. Later, Saudi women ran them – and were allowed to drive for the first time, too. Predictions that there would be a spate of car accidents didn't materialise. As in most countries, women are safer drivers than men. Many immigrants left the country following the liberalisation of women's rights as their jobs were replaced by Saudi women. When I visited in 2018 there had been a major change in the country. No more aggressive reception on arrival at the airport. I was respectfully welcomed when I arrived at passport control.

The stores were being managed efficiently by Saudi women and the religious police were no longer in evidence. The whole atmosphere was different. Even in 2004 there was a marked difference between life in Riyadh and Jeddah, the second largest city in Saudi Arabia, which is on the Red Sea. Jeddah was more informal and relaxed with fewer women wearing burkas.

The Alhokair brothers were very hospitable. I was invited to their large compound, where each of the brothers and the close family had houses, to celebrate the ending of the Ramadan fast. I was sitting with Paul, an Englishman who managed their franchise business, waiting for the family to return from prayers. It was a room with a large rectangular table set for all the family and a large television was broadcasting what looked like a Saudi soap. I was invited to sit next to Fawaz, as guest of honour, and was made to feel very welcome. The meal started with dipping a date into a glass of water, to signify the end of the fast and then a delicious meal of various Saudi delicacies including *kabsa*, a combination of meat, vegetables and potatoes flavoured with spices. I was struck by how important it is in most cultures and religions for the family to get together to share a meal, particularly on special occasions. It reminded me of my family getting together to celebrate a Jewish festival.

Unfortunately, we were not successful in Saudi Arabia. After a few years the partnership with Alhokair ended. The lack of success was mainly down to us. We didn't have the resources or systems to manage the business effectively. We hadn't appreciated the differences in climate and culture that meant that a lot of our range wasn't suitable for the market. The size of the business wasn't big enough to adapt the range for the Saudi market. As far as Alhokair was concerned there wasn't enough potential for them to invest in us. Their focus was on brands with greater potential like Zara. Alhokair has grown into a multibillion dollar business with an emphasis on developing and investing in shopping centres and other properties.

We had learnt an important lesson from our experience in Saudi Arabia. We put in place a team that could focus their efforts exclusively on our international business, and particularly the Middle East. We had an outstandingly good occasion wear range, and the Middle Eastern customer like dressing up and wearing shoes with plenty of diamantes and bling, so there was a natural synergy.

We were much more successful in 2008 with our second partner in the Middle East, Apparel. Apparel, benefiting from the entrepreneurial flair of Apparel's chairman Nilesh Ved and his wife Sima, was established in 1996 and was a successful retail conglomerate with a large stable of brands that it managed in the Middle East, India and other markets. It was already strong in the footwear sector, with distribution arrangements with Aldo, Nine West, the large US footwear brand, and at the time one of the largest footwear companies in the world, and the Singapore-based handbag and footwear company Charles & Keith, that was strong in bags and expanded successfully in Asian markets, in particular China. Apparel later added Skechers, Steve Madden and many others to its portfolio of footwear and accessory brands. The fact that it specialised in footwear and handled a lot of footwear brands could have been considered a negative as we would have a lot of competition. In fact, it was a positive because to be successful we had to have a point of difference to be able to compete with these brands in the market.

Fortunately, our range sold well, especially the dressy styles. One of the key requirements when entering a new market is to have the right partner. In many of the cases where we were not successful it was because we chose the wrong partner. We were fortunate with Apparel, which proved to be an excellent partner. One of its key qualities was the relationships it had built with the shopping mall owners. Getting the right stores in the right location in the key shopping centres is crucial and competition for the space is intense. Apparel had a wide range of major retail brands which gave it a strong negotiating position with the shopping centres. In addition, Nilesh Ved had built important personal relationships with many of the mall owners who respected his business acumen and commitment and were therefore sympathetic to his request for stores. The great thing about Apparel and Nilesh was that if you were successful, they would aggressively open new stores. Many partners "park" a brand after opening a few stores. That was not Apparel's philosophy. If a brand was successful, Apparel opened more stores. As a result, Dune stores were opened in all the major shopping centres in Dubai, Abu Dhabi, Kuwait, Qatar, Bahrain and Saudi Arabia.

The test of a relationship is when things go wrong, which they inevitably do. It is the ability to work collaboratively to solve the problem that defines a strong relationship. I remember after a couple of bad seasons Nilesh asked me to fly to Apparel's head office in Dubai. He produced a graph which showed Dune's performance relative to their other footwear brands. It wasn't pleasant viewing. The others were going steadily north. We were drifting south. "How are you going to solve this, Daniel?" was his question. We solved it by moving from a "pull" to a "push" model. A "pull" model is where the range and the depth of the buy is selected by the franchisee, in this case Apparel. A "push" model is where the franchiser (in this case, us) buys the range and decides how many pairs we should buy of each style. We knew the range much better than Apparel. We were better placed to buy the right styles. And so it proved. Our performance improved and the line on the graph started to head north. Of course, we worked closely with Apparel's team and listened to their feedback on the range. As we were controlling the buy, we also guaranteed a certain level of profit margin. After that change we grew from strength to strength.

Not all markets were as successful as the Middle East. We launched Dune in China with an experienced partner based in Hong Kong who had worked for many years as the distributor of Nine West in China. We opened concessions in department stores in some of the major cities, including the large Sogo department store in Hong Kong. We had limited success, but the partner had financial difficulties which made it impossible to continue.

I visited potential partners in Wenzhou and Nanjing. Wenzhou was on the east coast and was a large port city. Apart from being a centre of the manufacture of men's shoes it was famous for two unrelated facts, the large number of emigrants to Europe (90 per cent of the Chinese immigrants in Tuscany, Italy are from Wenzhou) and New York City, and being a centre of feng shui in China. I was met at the airport by a smartly dressed chauffeur in a limousine and driven at speed through the town to meet the chairman. The feng shui wasn't immediately apparent as Wenzhou, to my untrained eye, looked like many of the large Chinese cities I had visited with a plethora of skyscrapers dotted across the skyline and large volumes of noisy traffic emitting vast plumes of exhaust.

We arrived at an impressive office building in the centre of the city, and I was ushered in to see the chairman in his large palatial office. He was an eminent-looking man who was interested to hear about the footwear market in the UK and the history of Dune. He explained that the company, Aokang, specialised in men's shoes and had shops selling competitively priced shoes across China. He saw Dune as an opportunity to elevate its offer and join a few higher quality more fashionable brands that it was adding to its stable. This operation was based in Shanghai, and he suggested I visit there to discuss a proposal with the team. I flew to Shanghai and met the team, which consisted of one man in an office. It became apparent that the project was at an early stage of development and unsuitable for Dune.

The company I saw in Nanjing was called Sanpower which had bought House of Fraser, the UK department store, in 2014 for £450m. It owned a large department store in Nanjing and proposed opening a House of Fraser to showcase British products. It wanted Dune to run the footwear department.

The chairman of Sanpower, a man called Yuan Yafei, had built a successful conglomerate which ranged from technology to retail. He was very hands-on and demanding and insisted that the House of Fraser team fly to Nanjing ten times a year for meetings. His office put the chairman of Aokang's office in the shade. It was extremely large, the size of a generous one-bedroom flat, with expensive antique furniture and a Chinese painting that I was told was worth many millions of dollars. I was ushered into his office, and I could see him in the distance through a fog of cigarette smoke. Chairman Yafei was a chain smoker who had special, very expensive gold leaf cigarettes made for him. The meeting lasted a few minutes during which we exchanged pleasantries, and he gave me a small gift of one of their technology products which I never got to work. He had his own private lift and exit from the building where his Rolls-Royce was waiting for him. The economics of the proposal to operate the footwear concession didn't work. It meant a lot of work for a very modest reward.

I never found a credible partner in China. It is a challenging market even if you have a good partner. The size of the country means there are different regions, climates and tastes so it is often best to start in a particular region. There are a few companies, like Belle and Stella, that have a vertical business model in that they manufacture the shoes and sell them in their own stores. This gives them an advantage both in the margin they make and the speed that they can get the product to market. To be successful you need a strong brand, a distinctive product and an in-depth understanding of the market, that is apart from the need to adjust sizes to fit the Chinese customers' feet which are typically smaller than those of UK customers. Social channels like WeChat are very important to engage the customer, so working with a partner who understands these channels and is prepared to invest in influencers and brand marketing to raise the profile of the brand is essential.

Russia was another market where our partner wasn't good enough, although given subsequent events that may have been a blessing in disguise. They didn't have the resources, contacts or knowledge of the market. They opened three shops as a trial, but the sales were disappointing. The main challenge in Russia, apart from finding the right partner and getting paid, was the climate. Given the cold weather customers wanted their boots to be fur-lined, preferably with real fur. The size of our business in the market wasn't sufficient to develop special boots. Russian women did like dressing up and wearing flamboyant shoes with lots of bling, so we should have found a niche in that market, but it never happened.

Price was another reason for our lack of success. For many of the less affluent countries where we opened stores our prices were too high to justify more than a couple of stores in the best locations. Two stores didn't justify the investment in time and resources to establish and manage the partnership. This was the case in Vietnam, Malaysia and Thailand. Although they weren't successful, they did teach us a lesson about concentrating on larger markets. As an exception to this we had surprising success in Libya, Pakistan and Nigeria. We only had two stores in Libya but despite the unstable political situation, we traded exceptionally well, that is except when they had to close because of nearby hostilities. The good thing about those three countries is that they bought the same range as we sold in the Middle East which made supplying them with products relatively easy. Pakistan and Nigeria had great potential. However the difficulty in getting paid, due to currency restrictions in getting money out of the country, limited the growth opportunity.

22. Buy Shoe Studio and triple in size

During the early 2000s my focus was shifting firmly from Browning to Dune, from importer to retailer. Dune's success meant we were getting valuable information from our retail operation of the styles that were selling well, but we were cautious (maybe too cautious) to use this information for Browning as we felt that it would cannibalise the success we were having in Dune.

To continue to grow Browning Enterprises we expanded the range into other categories, mainly at the lower end of the market with products such as men's synthetic footwear and cheaper casual products that appealed to the supermarkets, like Tesco, and discounters, like Primark. Going downmarket had limited success. We were not adding enough value, and we couldn't compete on price with many of the lean operations.

At the end of 2007 we tried to sell the business so we could concentrate on the retail business. We appointed BDO Stoy Hayward to market the company. The timing wasn't good. The Bank of England had to support Northern Rock, a medium sized and highly leveraged bank, which led to a run on the banks, so confidence was at a low ebb. We had also had a difficult 2006 following the imposition by the EU of heavy duties on leather shoes from Vietnam and China when our profits fell from £3.5m to £2m. Unsurprisingly we received no offers.

The final straw and what persuaded me to close Browning was a visit I made to Next in 2009, then our largest customer. Next had placed a large order for a shoe but the price it was offering gave us a very small margin. At the time I was not very involved in the business. Duncan Miller, the CEO, and Susannah Huller, the product director, had come to me to get my opinion on whether we should take the order. The price it was offering was £1 lower than an equivalent order it had placed the previous season. For all the effort and risks I didn't think it was worth it. We didn't want to place it in a cheaper factory and compromise on the quality. I decided to visit Next with Susannah to meet with the buying director to try to persuade her to increase the price. The visit was not a success.

The buying director was charming and after hearing my argument asked me to wait in the meeting room. She returned five minutes later with our sample and another one that was almost identical. She was being offered this identical shoe by another supplier at 50p below the price she was offering us. Because of the important relationship Next had with us she was sacrificing the 50p. I left Next with the sense that Browning was going to struggle to survive in the face of this competition. We had a stark choice. We had to make drastic changes to our structure or wind down the business. In the end, in 2009, after 23 years of trading, we chose the latter course of action. Maxgreat continued to sell successfully in the UK market. It went directly to the retailer with Susannah and another ex-Browning employee, Brigit Turnbull, as the agents.

But as sometimes happens, in life as well as business, as one door closes, another opens. As an entrepreneur, I was always alert to opportunity. While this could sometimes risk distracting the business, as in the case of the Aldo franchise we decided not to take on, other opportunities were worth exploiting.

One came up in 2009, when things were already very busy. We did not only close Browning Enterprises. By coincidence, after 30 years in our small, terraced house in Browning Close, Maida Vale, West London, we had moved to a flat in Mayfair which had stretched our finances . I was still working hard but travelling less. Travel was more to visit our stores than overseas suppliers.

Over the past 17 years we had established Dune as an important player in the upper mid-market footwear space. Sales had reached £50m and we were making a modest profit of £2m. We had a portfolio of 45 stores in prime locations in the UK, 35 concessions which were performing well, an expanding website, which we had set up in 2005, and a nascent international franchise business.

The question was: what next? I was now focusing all my energies on Dune. Although in my sixties I was still ambitious. My goal of Dune becoming the leading affordable luxury footwear and accessories brand in the UK and globally remained. The question we faced in 2009 was how to grow Dune. Opening more stores wasn't an attractive option as the return from the investment was disappointing. The combination of high rents and rates made it challenging for a single-branded footwear retailer like Dune to make an acceptable profit. Online sales were growing but not fast enough to compensate for the disappointing profit generated by the stores. The most attractive growth opportunity was through third-party sales which in the UK meant concessions in department stores. Despite the high commission rates, we were making a reasonable return on our concessions in John Lewis. Our approaches to the other department stores had been rebuffed, so there was no immediate opportunity of opening concessions while Shoe Studio and Kurt Geiger monopolised the distribution.

The other area of growth, apart from international franchise, was wholesale. However, we were set up more as a retailer than a wholesaler. We had to narrow our range for the wholesale market and make a stronger brand statement. The wholesale range we were offering to customers was too wide and confusing. It also had to be ready by critical dates. If we missed these dates, then it was difficult to get customers to place orders. This conflicted with our strategy as a retailer where there was an advantage to placing orders later. The later you placed orders the more information you had about the direction of fashion. This could result in a more successful range. We had to change the way we built and bought the range if we were to become a successful wholesaler, something we belatedly did some years later when we built one global range that we sold through all our distribution channels.

Our growth ambitions were tempered by the dire economic situation. The subprime mortgage crisis of 2008 in the US triggered a severe recession which had devastating consequences for the global economy. Banks had been lending irresponsibly to households whose properties were worth a lot less than the amount of the loan. The banks themselves hadn't built the reserves necessary to deal with the flood of defaults. Banks like Lehman Brothers in the US collapsed and many financial institutions in the West either failed or had to be rescued by their governments. In 2009, the UK's GDP fell at the fastest rate since 1931 and unemployment jumped from 5 per cent to 10 per cent. Many households were highly geared and couldn't repay their mortgages which meant that consumer spending plummeted. The UK economy didn't fully recover until 2016.

Against this backdrop, an opportunity arose to buy Shoe Studio. The company had been borne out of the concessions operated by the British Shoe Corporation in department stores. David Spitz, a South African footwear entrepreneur, and Don McCarthy had acquired the concessions in 1991 and consolidated them into Shoe Studio. The company traded successfully and built sales to £100m. Don McCarthy, who was a successful entrepreneur and philanthropist, went on to become executive chairman and CEO of House of Fraser. The company was sold in 2006 to a clothing group called Mosaic, owned by an Icelandic company. Shoe Studio and Kurt Geiger controlled the footwear concessions in department stores. Buying Shoe Studio was a massive opportunity for Dune. Mosaic's CEO was Derek Lovelock, a highly respected and experienced clothing executive. I got to know Derek when we had worked with Oasis (one of Mosaic's brands) on its footwear range. It hadn't been a great success, but that was mainly due to the small footwear area rather than the quality. I spoke to Derek about buying Shoe Studio as it didn't fit into a group that was primarily a clothing company. There was little focus on Shoe Studio, whose performance was deteriorating. However, the price that Derek wanted for the company was too high. Meanwhile the owner of Mosaic, an Icelandic company called Baugur, was in trouble and went into administration. The knock-on effect was that Shoe Studio went into administration.

Here was an opportunity to become a major player in department store concessions in one fell swoop. Shoe Studio had concessions in Selfridges, House of Fraser, Fenwicks, Debenhams and Topshop, as well as in the top department stores in Holland and Denmark. Apart from the over 300 concessions there were 13 stores, a small website, an exclusive distribution agreement with Nine West and a stable of brands which included Bertie, Pied a Terre, Roland Cartier, Roberto Vianni and Chelsea Cobbler. The company had sales of over £100m. If we bought Shoe Studio we would treble our sales. Most importantly we could reduce our dependence on our stores and substantially increase our profits.

In 2008 John Egan had joined Dune as CEO. John had been retail director of Shoe Studio for eight years. He was a close friend and colleague of Don McCarthy. John was an ebullient character and outstanding retailer who brought a wealth of experience as well as great energy to the company. His knowledge of Shoe Studio was immensely valuable in understanding the business. He knew both the team running the business as well as the key personnel at the department stores.

The process of buying a company is a lengthy and painful one. The company was in administration which meant that we were dealing with the accountants, Deloitte, who had been appointed administrators. In January we received the information memorandum which was a 70-page document describing the business and giving details of the financial performance. In the current financial year, the company was forecast to lose £5m. The information memorandum is a selling document which tries to show how the business can be turned round and become profitable. We weren't interested in buying the business as a going concern. The attraction to us was acquiring the concessions, the brands and the stock at an attractive price. We would get substantial synergies from merging Shoe Studio's operations into our own. We wanted to take on as few of Shoe Studio's head office costs as possible. Unfortunately, that meant making about 80 people redundant. However, we were saving a lot of retail jobs by keeping the concessions.

Having done our due diligence we concluded that at the right price this could be a very attractive acquisition. Our first port of call was the CEOs of the department stores as we needed their approval. Shoe Studio had an existing concession agreement with each of the stores which was an attractive one. On going into administration, the agreement was automatically terminated. We wanted to continue the agreement in Dune's name. There was the risk that either they wouldn't approve of us taking over the Shoe Studio agreement or they would use it as an opportunity to increase the commission rate. We were dealing with three experienced negotiators. Selfridges' CEO was Paul Kelly, an imposing Irishman with a long and distinguished record in retail. Rob Templeman of Debenhams was renowned for being a tough negotiator, and of course Don McCarthy, who, although a good friend of John Egan and sympathetic to Dune, was known for his astute deal-making. Despite the friendship he was going to negotiate the best deal he could. Our meetings with them went well. They seemed on board with the principle of Dune replacing Shoe Studio, although they stressed the need to offer not only the Dune brand but also all the external brands that Shoe Studio was offering. We didn't have any experience of managing external brands, but John was confident that the team at Shoe Studio that managed external brands would join us and bring their expertise with them. There were rumours that Debenhams was supporting a small private equity company called Argyll, which we understood was also bidding for Shoe Studio, but we never found out if this was true. After the meeting Don McCarthy phoned me. He would support our acquisition, but he wanted an agreement that House of Fraser would have an option to own ten per cent of Dune when Dune was sold. His reasoning was that House of Fraser was the biggest partner of Shoe Studio, it was a very lucrative partnership, and it was reasonable that House of Fraser would benefit if I sold Dune. I wasn't happy giving away such a large amount of the equity. I tried to negotiate the percentage down to five per cent but unsuccessfully. Reluctantly I signed the agreement. Whether I should have accepted Don's proposal is questionable. Our negotiating position with the department stores was strong. One of their two footwear concessionaires had gone. Did they really want to put all their eggs in the Kurt Geiger basket? With two partners they could play one off against the other. With one there was a danger of their negotiating position being weakened. There was no point regretting the decision. The deal had been done.

As I have mentioned, footwear is a difficult category. Apart from the high stock holding due to the number of sizes, it needs shop staff to sell them. Department stores want to keep their staff costs to a minimum so there is a reluctance to staff the footwear department, properly, which often means lost sales as the customer is not prepared to search for a sales assistant to go to the stockroom and get the shoes. We estimate that the sales of footwear in a department store will increase by a minimum of 20 per cent if it is properly staffed. Although department stores could operate their own footwear department as they did in the US and many European countries, the combination of the staffing cost, the stock holding and the lack of footwear expertise persuaded them to stick with the concession model.

Now that we had the department stores on side, we needed to decide which parts of Shoe Studio we wanted to retain. The 13 shops were losing money, so it was an easy decision not to bid for them. Nine West was an important part of its range. We felt that Dune was a similar but stronger brand to Nine West, but we were much more attuned to the UK market, so we decided that we would end the distribution agreement with them. Their top team flew from New York to try and persuade us to change our mind, but despite their protestations we weren't persuaded. We decided against taking over Shoe Studio's website which took less than one per cent of its sales. We would incorporate its ranges onto our own website. Shoe Studio had a few luxury concessions where it was offering Manolo Blahnik, Prada and other top end brands. We decided not to take on these concessions. It was outside our comfort zone and took us too far away from core ranges. In addition, Shoe Studio was holding a lot of slow-moving stock and generally did not trade well.

Visiting Shoe Studio's head office to talk to staff was a sensitive issue as many of them were going to lose their jobs. There was a sense of sadness as people who had worked together for many years and built strong relationships were going their different ways. We did employ some of their head office team to fill gaps in our structure and to help us with the integration of Shoe Studio's operations into ours.

An important feature about buying a company out of administration is that you are buying the assets. You do not want to take on the liabilities. That was a difficult concept to explain to suppliers who were hoping that we would pay them the amount they were due. After the administration is finished, if there is anything left over after paying the administrator and the preferred creditors (like the banks), this amount is divided pro rata among the unsecured creditors. It was unlikely in this situation that anything would be available for unsecured creditors. The foreign suppliers were particularly aggrieved. They didn't understand the concept that we were only buying the assets. I had some difficult conversations explaining that this was not our responsibility. I remember one emotional Neapolitan factory owner pleading with me to pay him. In the end we set aside an amount to pay some of the most deserving suppliers.

The main item we were buying was the stock which was in the balance sheet at a cost of £20m. We pored through the stock sheets to try and assess its true value. Most of it was reasonably recent, although many of the styles did not have the full range of sizes. The question was how much to discount the cost of £20m. Reducing the value by 20 per cent was reasonable but often in a distressed sale situation the value is reduced by 50 per cent. The other items we were buying were the brands and the concession fixtures. We had to decide what to offer for the business. Apart from Argyll we weren't aware of any other bidders. Unless you were in footwear, or mistakenly wanted to get into footwear, Shoe Studio wasn't an attractive buy. After several meetings we made an offer of £4m. To our amazement it was accepted. It was a fantastic deal. That year our sales were £140m and we made a profit of £7m.

23. Banking problems

The acquisition of Shoe Studio transformed Dune. We only had one problem. How to pay the £4m we owed the administrator? We didn't have £4m in free cash. Despite demonstrating what a good deal it was and how we would sell the stock and pay back our bank within six months, it wouldn't lend us the money. Following the financial crisis banks were being very cautious. A lot of them had their own liquidity problems which restricted their ability to lend, even on such a compelling investment. I went to my personal bank and offered a charge on my flat as security. It was sorry, but it couldn't help. Banks can be frustrating. They often seem very set in their ways. One of the criticisms of UK banks, unlike banks in the US, is that they are too risk-averse. This is one reason why innovative businesses often find it difficult to raise money in the UK. Not having the funds was a problem. It would be highly damaging and embarrassing to have to go to the administrator to tell him that we didn't have the money to pay him. In the end, after calling in lots of favours from suppliers and partners, especially House of Fraser, we managed to delay payments and trade the stock to generate the cash internally to pay the administrator.

The year after the Shoe Studio acquisition in 2010/11 produced a strong profit. We had acquired a lot of stock at a very attractive price. We were able to sell it and make a healthy margin. The stock package was fragmented but the quality was acceptable, and it was reasonably current. The department stores gave us permission to have sale racks in our concessions to dispose of the stock. They benefited as we sold a lot of stock, and they made a good commission on the sales.

The integration of the teams had gone smoothly. We limited the number of Shoe Studio head office staff that joined the Dune team to the minimum. We wanted to keep our overheads low. We now had a much bigger business to manage. Initially we had decided to pull out of Debenhams but after negotiations with Rob Templeman and his successor, Michael Sharpe, we opened Dune concessions in September 2009 offering a small Dune range and a larger range of the sub brands that we have bought from Shoe Studio like Roberto Vianni and Roland Cartier. In the following years we also developed a brand we owned, Head over Heels, as a cheaper version of Dune to appeal to the less affluent customer at many of the Debenhams and some of the House of Fraser stores.

It became apparent in 2010 and 2011 that we didn't have a strong enough team to manage the enlarged company. In addition, our computer systems had been creaking even before the acquisition. Our main ERP – stock control – system was buckling under the pressure. We now had to consolidate all the Shoe Studio stock onto our system, making a difficult situation worse. In 2011 we hurriedly went ahead with the adoption of a new stock control system called Island Pacific. One day we had the old system. The next we had the new one. It was a disaster. We completely lost control of our stock which meant allocations and replenishments were done blind. It took months to sort out and, in the meantime, we lost a lot of money. It was an important lesson in planning and executing major projects. As a result of the system failure, we went from a strong profit in 2010 to a small loss in 2011.

After buying Shoe Studio we decided to change banks. We wanted a bank that had a specialist retail team. We were offered a generous bank facility by both Barclays, Shoe Studio's previous bankers, and Lloyds. Lloyds' offer was marginally more attractive, so we went with Lloyds, whose motif was a galloping black horse. I hadn't had a banking relationship with Lloyds on either a personal or business basis.

When we made Lloyds aware that we were having problems and were likely to breach our covenants, and make a small loss, they took two actions. Firstly, we were transferred to the "bad" bank that looks after problem accounts that had breached covenants. The manager, a retail specialist with a strong reputation, who had negotiated the facility and was one of the reasons we moved to Lloyds, suddenly disappeared. We never saw him again. In his place we were given a new manager who was aggressive and unsympathetic to our situation. His sole consideration was whether the bank had enough security for its loan. Secondly, Lloyds appointed a firm of accountants, BDO, to carry out an expensive report on the business to assess its viability. The report was inconclusive, although largely supportive of the management. I had a meeting with the Lloyds manager to discuss the report. He said if Lloyds were to continue to support the company, it would need more security. He would need a second charge on my home. I refused. It was in the joint names of my wife and I and my wife would not accept a second charge. Okay, but did I have any other security I could offer? What about my flat in France? Were there any other assets I could produce as security? I tried to remain calm. We had been trading since 1992, in the last ten years we had consistently made profits. Lloyds had unlimited charges on the whole of the business. No, I didn't have any other security. Its existing security would have to do. He wasn't happy. He would have to review the matter with his credit committee.

In the meantime, we had approached the US bank, Wells Fargo. I knew its UK CEO, a charming and astute banker called

Steven Chait. Wells Fargo provided what is called asset-backed lending. It looked at the value of our stock and debtors and based on that it offered a facility which was 50 per cent higher than the one given by Lloyds at a lower interest rate. The covenants were also much less demanding. I met with the Lloyds manager to tell him we were going to close the account and move to Wells Fargo. He said we were making a big mistake. He was willing to withdraw the demand for further security. We should stay with Lloyds, he said, because it was the better bank. He flipped from being the aggressive banker, insisting on me providing extra personal security, to a salesman selling me the benefits of Lloyds Bank. I have always managed to control my emotions. I rarely lose my cool. On this occasion I was close to the edge. I wanted to tell him where he could shove his facility. I resisted the urge and left the room. We moved to Wells Fargo where we stayed for the next five years. Our profits grew. The relationship with Wells Fargo was excellent and, importantly, stress-free. It is interesting how one bad experience can give you such a negative perception of a company or brand. We were just unlucky. The market was in turmoil and Lloyds was protecting its position. It could have handled the relationship better.

From 2010 we strengthened our management team. Our most important appointment was James Cox who joined as Finance Director from Thomas Pink, the shirt retailer. At the interview, at some brutalist 1960s office block in the City of London, he came across as a formal individual, wearing a dark suit and tie. He reminded me of our bankers. But he was impressive with a sharp mind and a lot of relevant retail experience. He was the best candidate, so we offered him the job. Our finance and operations teams had been struggling. The additional pressure from the acquisition had meant that our management reporting system was failing, and our operational systems were not fit for purpose. Over the next three years James transformed our operations. He employed a talented and experienced team. He put in place systems and a management information pack that enabled us to run the business in a professional and controlled manner. He built a strong relationship with our bankers. Giving them timely information, communicating any potential hiccups well in advance, was the key to building trust. Banks do not like hearing bad news after it has happened. If you warn them some time in advance, they are much more likely to support you in difficult times. That was certainly our experience, and it proved to be particularly relevant when the next big shock happened.

24. Marketing the brand

Between 2010 and 2020 our annual sales grew to £200m. After the acquisition of Shoe Studio, 65 per cent of the sales came from our concessions, which was an unhealthily large figure given where the department stores were heading over the following years. Our store sales didn't increase materially over the period. The main growth was in online sales which went from 7 per cent to 30 per cent of our turnover. We had set up the Dune website in 2005, but, like web retailing elsewhere, it was slow going at the time, before the invention of mobile phones which could surf the internet – which took off in the 2010s.

During this period there was a major investment in upgrading our website onto a new platform, adding functionality and making more of a brand statement. We also started spending serious amounts on performance marketing such as adverts on Google and Facebook and making sure Dune was at the top of Google's listing when you searched for the brand. At that time the whole social media space was exploding. The social network platforms increased their total user-base from 970 million in 2010 to 5.17 billion in 2024. So not only advertising on social channels but also creating and building Instagram and

Facebook accounts to engage customers and share brand stories and products became increasingly important.

The growth in online sales went hand in hand with our omnichannel strategy which aimed to give the customer a seamless experience both in store and online. The customer journey could start by browsing in the store and end by purchasing or researching online and buying in store. Omnichannel enabled the customer to choose how they wanted to engage with the brand, where they wanted to place their order and how they wanted the order fulfilled. The key to success, apart from having a compelling range at the right price, was to have a strong and consistent brand message through all channels, whether it is shops, website, social channels, press and advertising, plus having an efficient operational process that delivers fast, reliable and efficient customer service. This strategy has preoccupied us over the years and is an increasingly important element of successful retailing. It has been enhanced by the clever use of technology which has personalised the customer experience by curating ranges to their preferences and communicating with them in a relevant and interesting way.

After staff costs, marketing has been our biggest overhead, both brand and performance marketing. The great thing about performance marketing is that you can measure its effectiveness. There is so much data from Google and Meta (Facebook and Instagram) that you can accurately assess the results of what you spend. For a short while our head of performance marketing reported directly to me. It was fascinating but very technical. I can't say I understood it all but enough to grasp the key principles. For example, there are several metrics for measuring the results of performance marketing. Conversion rate (how many of the customers who visit the site buy the product) is a simple one but there is also CLV, CTR, ROAS, CPC, CAC, CPL. There is a danger of being blinded by science and acronyms. For me the key metric is return on ad spend (ROAS) which measures the revenue generated for each pound spent on advertising. We were aiming for a ROAS of 15 per cent which meant you spent £15,000 to generate £100,000 of sales. One thing was clear: the more traffic you could drive to your website through emails and low-cost tools such as search engine optimisation (SEO), which pushes you up the rankings when you search on Google for a brand or product by clever wording of the website and other media communications, the better. Once you use Google or Meta adverts it becomes a lot more expensive. Their costs have gone up year after year. Google is frequently upgrading its systems and algorithms, which makes life complicated, but no doubt helps to contribute to their enormous profits. The fact that Google has a near monopoly (92 per cent) of internet searches means that you have no choice but accept its changes even though they may add to your costs. New AI chatbots, like ChatGPT, are challenging Google's search monopoly, and no doubt new developments in this exciting area will provide more competition to Google.

Brand marketing is much more difficult to measure. A 19th Century US merchant said: "Half the money I spend on advertising is wasted; the trouble is I don't know which half." You need brand marketing to make people aware of the brand. The luxury brands spend untold fortunes on brand advertising. Every magazine or newspaper you pick up has pages and pages of the same few luxury brands. It is the frequency and prominence of these adverts which have such a strong impact. Using A-list celebrities and models wearing their clothes or carrying their bags make them very desirable and aspirational. This helps them make big profits by, for example, selling a bag that costs around £200 for upwards of £2,000. The luxury brands are finding it more difficult to sell their products, mainly because they have increased their prices to such a level (way above the rate of inflation) that customers baulk at paying them. At a certain level you feel ripped off.

Until 2011 our main marketing spend was on public relations (PR) and of course our shops, which were in prime locations. We had a small PR team who spent most of their time sending samples to the press so that they would be used in their editorial content. This was an effective and low-cost way of getting the brand in front of potential customers. I have always had a lot of stick from my wife, mother and mother-in-law over the years when there was a page of shoes or bags in a magazine and there was no Dune product featured. In 2011 we started our first print advertising campaign in magazines like *Elle* and the *Sunday Times Style*. This was partly to broaden the reach of the brand and attract new customers but was also a statement to our partners, like Selfridges, that we were serious about investing in the Dune brand. Having adverts in magazines also makes it easier to get editorial coverage and gives our PR team additional leverage. Over the years we have tried adverts on buses, taxis, in the underground and outdoor sites. I am not sure how effective they have been. My suspicion is not very effective. This is possibly because we haven't had a sustained campaign through these media for several years. This is the issue with brand advertising. Not only is it difficult to measure success but also you often don't see the results for several months or years after the campaign. Marketeers say that a customer needs to engage with a brand eight times (whether it is visiting the shops, the website, social channels, adverts etc.) before they think about buying, which may explain the difficulty in measuring the success of most brand marketing. One lesson I have learnt is that you can't chop and change your advertising media too often. You need a consistent strategy that you stick with for several seasons or years to get the real benefit.

One area of marketing that was to grow massively from 2010 is celebrity endorsement and the use of influencers. An important experience of the power of celebrity was when Kate Middleton, then Duchess of Cambridge, visited Canada in June 2011. This was two months after she married William so there was intense public interest in the couple and in particular Kate, and what she was wearing. The press was full of photos of her looking radiant and stylish. They were very complimentary about her dress sense. She was seen on several occasions wearing a pair of Pied a Terre espadrilles. We had acquired the Pied a Terre brand when we bought Shoe Studio in 2009. The press had found out that her shoes were from Pied a Terre. We were inundated in requests for the espadrille in the UK but also in Canada and the US. The style sold out within days. We continued to sell the style for several years. A large proportion of the sales were to the US. We were very lucky that Kate wore our shoes. It is not something you can plan. There is no doubt that getting well known people to wear your product makes a big difference to the sales. More recently we have spent less on traditional marketing like press adverts and more on celebrity placement. A recent photo of the actress Katie Holmes wearing our Deliberate bag, a soft, oversized woven bag, while walking in New York, had a dramatic impact on the sales both in the US and UK. We reposted the image on many of our social channels. The press picked up on it and the image was in several newspapers and fashion magazines. We sold out of the bag in a few days and had a long waiting list for the repeat orders. It is still our best-selling bag. It demonstrates the increasing selling power of celebrity.

With the launch of Instagram in 2010 we saw the gradual emergence of influencers as an important element of the marketing mix. What started out as a hobby has become a major income earner for successful influencers. Cristiano Ronaldo, the world's biggest influencer, has 645 million followers. The influencer industry today is worth $21bn. Influencers help brands build their customer base. Their followers trust them and buy their recommendations. Our experience with influencers has been mixed. Like everything else it's about having the right influencers who are authentic, and really like and engage with your brand. Too many influencers represent too many brands. Every day they are promoting a different brand. As a result, the relationship is transient. We are now much more selective. We only use influencers that relate to Dune and have followers who are potential Dune customers.

25. Building the Dune team

From 2010 our business became a lot more complicated. We had acquired new brands that needed a team to design, buy and merchandise the ranges. We were managing all the House of Fraser men's footwear departments, so we needed a team to choose the external brands and select the styles. We had social media channels to manage. We needed specialists in performance and brand marketing. As web sales grew, we needed bigger ecommerce and IT teams to facilitate the growth. Following the acquisition our sales had trebled but there was a danger that all the extra profit would be eaten up by the additional overheads. There was a tension between keeping our overheads as low as possible and the need to take on staff to manage a more complex business. We were selling through so many different channels, each one delivering a good contribution to profits, and each channel needed a team to support it. The danger was that we were running to stand still. We were making good profits, but they weren't increasing as fast as we wanted partly because our cost base kept on going up.

Our board of executive directors was led by John Egan who was CEO until 2017. His strength was building relationships with all our partners, particularly the department stores. He was an upbeat extrovert who developed a strong sense of corporate spirit within the business. James Cox had been promoted from CFO to COO. He looked after the finance, IT and logistics teams. His strength, apart from being an intelligent and thoughtful manager, was building an excellent team to manage these areas and putting in place an operational system that was very effective and reliable. He built a strong sense of loyalty with his team. James had hired Alice Arnold in 2010. She became CFO in 2010. Alice was a strong finance professional who was totally on top of the numbers and was greatly respected by the bank and our professional advisers.

I remained in charge of buying and merchandising. We were a product-led company and after the acquisition the variety of ranges made life much more complicated. I was aware that if we got the product wrong then it would be damaging. I had seen many fashion companies fail because their ranges were poor and not on brand. I wanted to ensure that this didn't happen to us. There were strengths and weaknesses in this structure. The strength was that we all worked hard and were good at our area of responsibility. We got on well and respected each other's expertise. The weakness was that it encouraged a siloed structure with teams not interacting as well as they should. It also, to some extent, undermined John and later James's position as CEO.

Debra Bloom joined in 2010 from a discount retailer called Select. Before Select she had worked for the Arcadia Group. Debra has great taste and fashion sense. She really understood the brand and was strong on the product. Having her in charge of design and buying made my life a lot easier. I was confident that our ranges would be on brand. Initially she was just responsible for women's product. She took over responsibility for men's product in 2019. When we acquired Shoe Studio the men's director, Justin Burzynski, joined us. As we now ran the House of Frasers men's departments and men's was an important and growing part of our business, we needed a senior men's expert.

Until Justin's appointment I had managed the men's buying team. After the acquisition of Shoe Studio, we had a lot of brands to manage so we needed a bigger and more experienced team to plan and buy the ranges

Nilesh Karia, our trading director, joined in 2005 from Faith Shoes. He was an exceptional merchandiser who managed the stock with skill and precision. He was commercial and entrepreneurial. He was fond of his WSSI (weekly sales, stock and intake) schedules, a sort of master schedule with detailed projections of sales and stock for the season. Because there were now so many different brands and ranges, I spent a lot of my time at range planning meetings and product reviews with Debra and Nilesh satisfying ourselves that the ranges were strong and on brand. After a while it was clear that I was too involved in the detail and gradually stepped back to allow the team more freedom.

The size of our concession business was a concern. The department stores were struggling as customers changed their buying habits. At some stage we needed to address this situation. Meanwhile, because of our focus on the department store business, we didn't devote enough attention to building our third-party wholesale business. Our retail business (stores, concessions and website) was such a large part of our business we hadn't focused enough attention on building relations with wholesale customers, especially international ones. It was only later when we made the necessary changes to the range, the processes and the team that we became a successful wholesaler.

In the summer of 2011, I got an invitation from the Prime Minister, David Cameron, to attend a reception at 10 Downing Street to celebrate the UK's entrepreneurs and fastest growing businesses. Following the acquisition of Shoe Studio our profits had shot up which was why I had been invited. I got a taxi to Whitehall where I passed through security to enter Downing Street which was a lot less impressive than it appeared on the television. It is a small cul-de-sac with a few terraced houses, the most prominent of which are Number 10, the residence of the Prime Minister, and number 11, the home of the Chancellor of the Exchequer. Once inside (after handing in my mobile phone) the interior was a lot more impressive with a large sweeping staircase with photos of all the past prime ministers. It was a glorious day, so the reception was held in the large garden. I mingled with the other guests, from an eclectic group of businesses. A young couple had started a recruitment company that had seen stellar growth. An Indian man had built a very successful business in car parts. There was a man who had a company that made the glue that stuck the fold on the top of a box of breakfast cereal. Some of these niche businesses made large profits. George Osborne, the Chancellor of the Exchequer, was wandering around. The government, a coalition of Conservatives and Liberal Democrats, was pursuing a policy of austerity to reduce the budget deficit. Maybe because of these cuts, there was a slightly downbeat atmosphere. David Cameron made a short speech congratulating us on our achievements and stressing the importance of entrepreneurs to the economy. I met him afterwards. He looked a bit distracted but asked me what I did. When I explained that I was a footwear retailer he looked at me sympathetically and said it was a shame that things were so challenging. In his opinion they were unlikely to improve any time soon. I suppose he was being honest, and his message reflected the times, but I would have preferred a more positive message from our Prime Minister. I left feeling more downbeat than when I arrived.

Although our store performance was mixed and sales were being driven by our online presence, we continued to invest in the key stores in the prime locations that were trading well. They were the beacons for the brand, and it was important they made a strong statement. We introduced a new shop fit in 2011. Although we liked the "shifting sands of Dune" concept, it was time for a change. Our shop designers, Four by Two, felt we needed a dramatic feature. They came up with the idea of a reverse catwalk down the length of the centre of the ceiling. We fitted a line of colourful high heeled stilettos on the catwalk (with the soles stuck to the catwalk). The overall effect was dramatic. People did stop and look. The performance of the refitted store improved so we rolled out the concept to our new and top stores. The only downside was that it made the store look more feminine when we were in the process of attracting more male customers. Putting men's shoes on the catwalk didn't have the same impact as a pair of sexy stilettos. Fashion was also getting more casual so sexy stilettos weren't the hot item they once were.

In 2014 I was awarded a Lifetime Achievement Award by the trade magazine *Drapers*. That year Dune also won the Multiple Footwear Award, so there was a double celebration. My mother attended the ceremony. She was clearly proud that I received the award. No doubt she was also surprised that, almost 40 years after my father's death, there was still a Rubin in the footwear trade.

26. Painful Lessons

My dream was to have a chain of Dune stores across all the major countries of the world. Very few British footwear brands have been able to do this. JD Sports and Sports Direct are two UK companies that have been successful in opening international stores, but they are billion-dollar businesses with substantial resources selling popular sports brands like Nike and Adidas. If you are a brand that has an innovative and distinctive product like Birkenstock or Ugg then there is also an opportunity for a limited number of stores to showcase the brand. But if you are like Dune, selling fashion shoes, albeit special ones, opening stores outside your home market is a challenge.

The lure of opening in the US was easy to understand. Here was a dynamic and exciting market that was ten times the size of the UK, and was recovering well from the 2008 financial crisis. Americans had similar fashion tastes to Brits and the fact that they almost spoke the same language made doing business a lot easier. If we could establish Dune – now branded Dune London – in the US, it would transform the business and increase its value. I had been to New York many times and liked the energy and positivity. Whenever I visited, I spent many hours walking round the shopping districts photographing shoes to get ideas and talking to the sales staff to understand what styles were selling well. I always felt that a Dune London store would do well in New York. The competition didn't look particularly daunting. I was also encouraged because a lot of US footwear companies visited our stores in London and bought our shoes for inspiration. They liked the range and felt it offered a different aesthetic to the US brands, more stylish and contemporary.

I was also encouraged because in 2013 we had received a tentative offer from Steve Madden to buy Dune at an attractive price. Part of the deal was that we would distribute Steve Madden shoes in the UK and possibly open stores. The fact that Steve Madden himself, who was red hot on product and was the driving force behind the company, liked what we were doing was a great endorsement. At the time we had a particularly distinctive range of dressy products which told a compelling and cohesive story. Steve Madden was less keen to open Dune stores in the US which made the deal less attractive as we would be devoting a lot of our energies on its brand in the UK but without any reciprocal arrangement in the US. In the end I decided not to go ahead but to focus on Dune London and consider opening stores in the US ourselves. With hindsight, this was a grave mistake.

Over the years I had visited Steve Madden's showroom during the New York shoe fair called FFANY (Fashion Footwear Association of New York) in February and August. Steve Madden was the dominant young fashion footwear brand in the US. There was always a buzz with all the major players in the footwear market milling around the showroom. The Steve Madden team had a single-minded commitment to selling shoes. Steve Madden himself would hover round the edge of the action, wearing his signature T-shirt, jeans and baseball cap, taking it all in. After spending some time working for a shoe wholesaler and learning the trade, Steve had started the company in 1990 selling shoes from the boot of his car in New York. He was a consummate trader and the drive, determination and resilience that he showed in building the company was reflected in the way the business operated and was a key factor in its huge success.

The range was dominated by some key styles. The US being such a large market it meant that those hero products would generate a large amount of their sales. The Steve Madden team were experts in developing these key styles and selling them widely throughout the trade. They would produce versions at different grades depending on the level of the retailer they were selling to. The high-end department stores would get the premium version and the discount stores would get a synthetic product that had been engineered to fit into the right price.

Even though the set-up was impressive and there was an energy in the Steve Madden business I felt that we offered something new and different. I therefore started looking for a shop to kick-start our US adventure. Looking at the competitors, Steve Madden and Aldo were strong players. Their prices were about 20 per cent below ours but our quality was better and our styling was different, a bit older and more sophisticated. We aligned ourselves with brands like Sam Edelman and Vince Camuto who were more expensive and aspirational. Interestingly neither had many stores. They traded mainly through department stores and wholesale accounts which was a fact that I should have paid more attention to. Maybe there was a reason they didn't have a lot of stores. After several visits we chose a prominent store in the main street in Soho, a major shopping destination. Many of the niche brands had stores off the main road in side streets, which were attractive but had a lot less customer traffic. As a new brand that was not well known in the market, we needed to be visible. We didn't want to fail because we were in a location with low footfall.

We entered the store at the top of the market. Whether the decision to take the store was driven by hubris or naivety I am not sure. Whatever it was, we would have to sell a lot of shoes to pay the rent. In addition, the cost of fitting out the store was high, and because we were obliged to carry out the project using union labour, who had strict work schedules, the whole project dragged on way past the agreed opening date and well over budget. I always thought that New York contractors would be faster and cheaper than in the UK. I was wrong on both accounts. We appointed a PR company to support the launch. We also had some good coverage on social channels and in the press, but despite this activity sales remained depressingly slow. We had found a red double-decker London bus which we drove round SoHo announcing our arrival and positioned a red phone box outside the store to emphasise the London connection. They were nice touches but in the big scheme of things had only a marginal impact.

Surprisingly our men's range sold well but that didn't compensate for the low sales of women's footwear. I realised that we were never going to reach the levels of sales we needed to break even. Reluctantly we closed both the Soho store and another one we had opened in a shopping mall on Long Island. What did we learn? Firstly, to be realistic about the sales you can achieve and whether, given the amount of rent you are paying and the staff and other costs, you will take enough money to make a reasonable profit. That is basic retailing but in the massive and exciting US market I got carried away and put unrealistic forecasts into the business plan. Secondly, it can take at least a year for customers to know that you are there, so you need to be patient. Consumers are conservative and it takes time for them to embrace a new brand. Unfortunately, with the SoHo store the rent was so high that even if we waited many years, we would never have generated enough sales to make the store profitable. Thirdly, you need to spend money on marketing. Dune London wasn't known in the US. We spent a bit, but not enough to move the dial. Getting a celebrity endorsement makes a big difference and combined with having a small group of the right influencers can change people's perception of the brand. Lastly, we failed to recognise how promotional the US market is. No one wanted to pay full price unless the product was special and different.

We needed to buy extra-high margin products so that we could promote when everyone else did and still make a good return.

In tandem with the store opening, we had opened a showroom for our wholesale business. We had appointed a salesman to drive our wholesale business. We did have some success with Nordstrom, a large department store based in Seattle. It sold our dressier styles well, especially a flat multicoloured jewelled sandal called Kylie. It was Nordstrom stores in the southern states that drove the sales which proved to be valuable information when we successfully re-entered the US market eight years later. With the other department stores we were less successful. We got caught by the "give back" system. The effect of this process was that at the end of the season you had the option of taking back unsold stock or allowing the department store to reduce the price of the stock that they were holding to a level where it sold. They then sent you a debit note for the cost of the markdown. We chose the latter option. I am not sure how it happened, but the debit notes from a group called Belk came to more than the value of the sales we made to them, which remains a mystery and doesn't reflect well on how we managed the business.

One of our main problems was that the buyers couldn't put us in any distinct category. They were confused as to what Dune stood for. Were we a dressy or a casual brand? The trouble was we were behaving like a retailer, not a brand, and were offering too wide a range. We should have narrowed it right down and focused on our dressy product, which was special and different from the competition. When we entered the US for the second time, we had learnt that lesson and had a much more distinctive and cohesive range. We left the US with our tail between our legs. It had been a humbling and expensive experience.

Switzerland was another chastening experience. In 2015, I was introduced to François Rueff, a successful businessman, based in Basel. François had built his own fashion retail company as well as successfully introducing several French and UK brands into the Swiss market. François was a softly spoken, serious man with a warm and generous personality. We already had two successful concessions in Geneva and Zurich with the department store group, Manor, so going into a joint venture with François to expand the number of concessions and selectively opening stores looked like an attractive proposal. There seemed to be an opportunity for Dune to occupy the middle market between the cheap operators and the high-end retail brands. If we were successful in Manor, then there was a logic in us opening a small number of stores in the top eight or so cities in Switzerland. We had an experienced partner who knew Swiss fashion retail intimately, and understood which were the right locations for our stores. John, Debra and I had visited the market, did our due diligence and felt there was a good opportunity for Dune in Switzerland. John was particularly keen as he knew Switzerland well from his previous experience with Bally. What could go wrong?

We set up Dune Switzerland that was 50 per cent owned by Dune and 50 per cent by François. We would look after the product, store concept and brand and François would look after the stores, the staffing and relations with Manor, although we would work closely on all areas. François's knowledge of the Swiss market was invaluable although, as it turned out, we underestimated the difficulty of having footwear and accessory stores. Over a period of a couple of years we opened stores in Basel, Bern, Lucerne, Zurich, Lausanne and St Gallen, and later a second store in a prominent position in Zurich. They were attractive stores in good locations. At the same time, we opened more concessions in Manor for both men's and women's footwear and handbags.

Visiting these cities was a pleasure as they were very attractive towns with a lot of character. Unfortunately reading the sales figures was not so pleasurable. The sales were a long way short of our budgets, even though the budgets took into account that it would take time to build the business. The main issue was that the traffic into the stores was too low. There weren't enough customers to generate the sales we needed to break even. With hindsight the population of each of these cities was too small. In addition, like the US, Dune was not known in the market. We had opened stores in prime locations and hoped that this would offset the fact that we were new to the market, but the Swiss customer was conservative and slow to embrace a new brand. In addition, while François's clothing stores had a wide range of products, we were just selling footwear and handbags.

When I visited Switzerland, it was immediately clear what the problem was: I could spend the morning in a store without seeing a customer. It was hugely demotivating for the store team and expensive for us as store staff were paid a lot more in Switzerland than in the UK. I also noticed that another long-established better grade footwear chain had closed its stores. This should have been a positive as we hoped that shoppers would come to Dune for their footwear. Unfortunately, that didn't happen. It was an unhappy confirmation that we weren't alone in struggling to make a profit from shoe stores. We made some changes to the range which initially was too dressy, but even our best sellers, which were flying off the shelves in the UK and other countries, were selling slowly in Switzerland. We closed most of the stores and were left with the best stores in Zurich and Basel, which we have subsequently sold. Writing off the investment we had made was painful, but this was preferable to supporting stores that had little prospect of making a profit. It would also be a distraction spending management time and investment in stock on loss-making stores. There is always the temptation to devote too much time on loss-making stores to try and find ways of turning them round. In the end it is often best to take the pain and close them.

In the end we bought out François's shares and became the sole shareholder of Dune Switzerland which consisted of two break-even stores, several handbag concessions in Manor and some footwear concessions which Manor had outsourced to a local operator. Like the US it has been a salutary and expensive experience. I keep in touch with François. We exchange thoughts on the state of the world, football and the progress of our grandchildren. He no doubt regrets his involvement with footwear. He should have stuck to clothing. We should have resisted the temptation to open stores in Switzerland. Despite our due diligence, success in Manor and having an experienced partner we had failed. We won't rush to open stores in another European country any time soon.

27. International expansion

The other major area of growth, apart from e-commerce, was international sales. We added London to our name in 2013 to reflect our growing international business. London was our home, and we wanted to be associated with its reputation as a vibrant, fashionable city, full of history and culture. On my many international travels, everyone I met loved London. Living in the city we often take its many attractions for granted.

Our main international expansion was in the Middle East. Due to its rich oil and natural gas reserves, many countries in the region were seeing substantial growth. Its leaders were using their wealth to open their economies and invest more actively abroad. Dubai had expanded rapidly, becoming an important trading centre and an attractive tourist destination. This had led to a property boom with the construction of impressive skyscrapers, luxury hotels and apartment blocks. Dubai was one of the fastest growing cities in the world. Unfortunately, when the financial crisis struck the country was overextended. It needed its rich neighbour, Abu Dhabi, to bail it out with a $20bn loan. Since then, Dubai has shown amazing resilience. After it had weathered the storm of the financial crisis it has seen strong growth, becoming the destination for many businessmen because of its dynamic economy and low tax environment. The other Middle Eastern countries also invested in new shopping malls to attract both local and international consumers. Our partner, Apparel, continued to open Dune London stores in the region. Nilesh Ved, Apparel's CEO, recognised the potential in Saudi Arabia, a country with 35 million people and a young demographic. The ambitious opening and expansion of the economy by its leader, Crown Prince Mohammed bin Salman, meant that there was exciting potential for opening Dune London stores in the country.

Dune London has over 60 stores in the Middle East. As new shopping centres open, some of the older ones have lost their attraction which means an active opening and closing store programme. The market is undergoing a change. As with the rest of the world, fashion is becoming more casual. Although we sell a larger proportion of dressy shoes in the Middle East, there is a gradual shift to more smart casual footwear. Bags also sell very well in the region. In Saudi, bags represent half of our sales.

Our partner in India is Reliance Brands, part of the hugely successful Reliance Industries built by Mukesh Ambani, the richest person in Asia, worth a staggering $120bn. He made his fortune in several industries including petrochemicals and communications, founding Jio, the largest mobile network operator in India and the third largest in the world. Reliance opened Dune London stores in Delhi, Mumbai and other cities in India. The stores in Delhi and Mumbai traded well. The dressy styles were particularly strong during the wedding season which is a major event in the Indian social calendar in the winter months. The stores in other cities were less successful, mainly because our prices were too high for the local customers. We tried a cheaper range called D by Dune, but we had problems meeting the minimum quantities required by the factories. Whenever we were persuaded to produce a D by Dune range by Reliance, it always ended badly. The factories were reluctant to make the small orders. When they did make them there were issues with the strict labelling requirements of the Indian authorities. The orders often arrived late which made Reliance unhappy, cost us money in discounts and upset the factories because of the disruption it caused its production. It was a good lesson in not trying to please partners by agreeing to small orders of special styles.

In 2018 I received an invitation to the wedding of Nita and Mukesh Ambani's daughter, Isha. Indian weddings are renowned for their opulence. The wedding of the richest man in India was going to be a magnificent affair. There was a weekend at Udaipur, renowned for its lakes, beautiful palaces, mountains and tiger gardens. The wedding itself was at the Ambanis' home in Mumbai, Antilia, an amazing construction and one of the most expensive private residences in the world located on billionaires' road, Altamount Road. For the entertainment, Beyoncé was being flown in to give a special performance at a reputed cost of $4.7m. Unfortunately, we couldn't attend as my wife was in a dance competition in Paris which trumped the Ambani wedding.

My last trip to India was in 2019 to visit our stores and some of the factories manufacturing our shoes. Navin Balani, who oversees our brand in India, took me round our stores in Mumbai and Pune, a town 100 miles from Mumbai, and Delhi. There is no substitute for visiting stores. You can see how the brand is presented in the market. There is always some aspect of the store or the range that can be improved. On this trip we agreed that the brand imagery, a heavy men's boot, wasn't appropriate for the Indian market and needed changing.

An impressive new mall was being built by Reliance in Mumbai called Jio World Drive. Dune had a prominent unit in the mall. Darshan Mehta, the CEO of Reliance Retail (who tragically died in 2025 at the young age of 63), had seen our store in Dubai Mall, the largest and most prestigious mall in Dubai, and the second largest in the world. We had designed a special shop fit for this mall that was opulent and had the "wow" factor with a combination of white Carrara marble, gold finished frames and specially etched glass shelves. Dharshan was impressed with the design and wanted the same design for Jio World Drive. The store opened in the autumn of 2020. The timing wasn't ideal as it was in the middle of the pandemic, but the store, our flagship in the market, has since traded well. Trade has become more challenging since 2025 when the new requirements of the Bureau of Indian Standards meant that shoes from China were greatly restricted. It has been difficult to replicate the fine dressy product that we source from China in Indian factories.

After visiting Mumbai and Delhi I drove to Agra to meet one of our suppliers. Reliance arranged for a driver to take me. A new motorway was about to open which reduced the drive time by over an hour. The existing main road from Delhi to Agra was very congested and slow. Reliance used its influence to allow the driver to take this motorway even though it wasn't officially open. We arrived at the start of this impressive six-lane motorway when a security guard opened the barrier. There were no other cars on the road. It was a strange experience driving for miles without anyone in sight. The main problem was the driver, an elderly man with thick grey hair and handlebar moustache. It was hot and he seemed mesmerised by the endless road with no distractions. After a little while I could see in the mirror that his eyes were slowly closing. Before they finally closed, he jerked awake. They then started closing again. I didn't fancy having a car crash on a deserted road in the middle of nowhere, with no mobile reception, so I started coughing and shuffling around in the hope that this would keep him awake. It worked for a little while but then his eyes started to close again. With three hours still to go I asked him to stop the car. He thought I wanted to take a toilet break. I got out of the car and asked him to get out as well. He didn't understand, shrugged his shoulders and shook his head. He didn't need a toilet break. I spent a few minutes attempting to act out someone falling asleep while driving, and then waving my hand in disapproval. Eventually he got the message and smiled. The rest of the journey went without incident except that he didn't stop talking. As he spoke Hindi, I didn't understand a word he was saying which stopped me from falling asleep – but at least kept him awake.

Unlike Delhi and Mumbai, Agra looked much the same as when I was there 25 years previously. Cattle were still wandering along the roads. The main change was the polluted air that engulfed the city. In the three previous visits to Agra, I had never visited the Taj Mahal, one of the new seven wonders of the world. I was determined to see it this trip. I had arranged for a guide to take me there early in the morning before the crowds arrived. We saw it as the sun was rising. It was a magnificent sight with a long reflecting pool leading up to the highly decorated marble structure with a huge dome and pillars.

I am always fascinated by India. Maybe because it is so different from the UK. The people are hospitable, and I love the food. Despite the poverty there is an energy and optimism. I wish I had more time to explore other parts of the country. From a business perspective I left feeling as I usually did. Here was great potential for Dune London but I questioned whether we would ever fulfil that potential.

After the Middle East and India, most of the franchise markets we entered were small. We initially made the mistake of building bespoke ranges for each partner. Not only did it take up a lot of time and effort but we also often found that the partner wanted to make changes. Given the size of the markets, we finally decided to prepare a global range, which rightly got closer and closer to the UK range, with changes to reflect the climatic difference. The partners had to place their orders from this range. This strategy has greatly reduced the workload of our team and has worked much better for both us and our partners. We now have franchises in the Philippines, Chile, Australia, Serbia, Algeria, Malta, Libya, Pakistan and Nigeria as well as the stores in the Middle East and India. We have closed stores in Vietnam, Malaysia, Thailand, Russia, China, New Zealand and South Africa. Most of the closures have been because the partner ceased trading, our prices were too high, or it was clear that neither of us could make any money from the partnership. Our focus is now more on finding bigger markets where there is more potential for growth.

The franchise model doesn't work for us in Western Europe. This is mainly because the market is more competitive which doesn't allow the franchisee to make a sufficiently attractive margin. Much like in the UK specialist fashion footwear stores have been closing in Europe at an accelerating rate over the past ten years. The economics just don't work. The combination of high staff and occupancy costs, high stock levels and increasing competition from non-specialists, like Zara, or online players, like Zalando, has made the model unattractive. The opportunity in Europe is either concessions in department stores (although these are getting more limited), wholesale or selling through local online stores. Brexit has made selling in Europe more challenging due to the increased administrative burden and logistical costs.

However, we still have a presence in mainland Europe. When we bought Shoe Studio, we acquired five concessions in a Dutch department store called De Bijenkorf, which is now part of Selfridges. These have traded well both in store and online. Similarly, our business with Zalando, the largest and most successful online player in Europe, has grown. The potential for further growth through Zalando and other online companies and department stores is a big opportunity for the business. We recently joined a distribution hub based in Germany that works closely with Zalando. This has helped avoid a lot of the red tape of shipping into Europe and opens a huge potential market which we have not yet exploited.

28. Pandemic pain

As we approached 2020, we decided to invest more in raising the profile of the brand. We appointed the actor Gillian Anderson to be the face of the brand for 2020. We chose Gillian as we wanted someone who wasn't just a celebrity, who hadn't been in lots of other advertising campaigns, but was an authentic and exceptionally talented actor, based in London. She had got the role of Margaret Thatcher for the TV series of *The Crown* which was due to be released in autumn 2020, which would greatly raise her profile. We chose Rankin, the photographer, to direct the campaign, having successfully worked with him a few years previously. It was our most expensive and ambitious campaign by a long shot and a big investment and we had high hopes it would raise the profile of the brand.

We created three sets reproducing different London locations: a flower market, a greasy spoon cafe and a stately home. Gillian's preference was to wear heels rather than casual shoes. We were pleased with the results. Gillian Anderson had a lot of character and looked terrific in the campaign. We had a lot to look forward to as a company, but then news started emerging of a mystery virus coming out of Wuhan in China.

The spread of the Covid-19 virus was a major news story, but it became real to me when my wife and I travelled to the Micam footwear fair in Milan on 15th February 2020. We had passed through customs when we were met by a long queue of passengers waiting to have their temperature checked by a group of white-coated officials wearing masks. We hadn't seen this type of response in the UK. There was clearly a major concern among the authorities about the spread of the virus. The first cases of coronavirus in Italy had been discovered on 31st January when two Chinese tourists from Hunan arrived in Rome. Lombardy was to become the worst coronavirus affected region in Italy with 50,000 deaths, more than double the next region. Once through the airport everyday life was continuing as normal. That evening we went to the opera at the famous Teatro alla Scala to see a performance of *Il Travatore*. It was packed. If there ever was a superspreading event, this was it.

The next day I went to the footwear fair in the vast Fiera Milano Exhibition Centre. It turned out to be my last visit. The next few fairs were cancelled because of the pandemic and even when it was over, the fairs became less important. Many of our suppliers stopped exhibiting, preferring to show at the earlier Garda fair. I had also lost some of my enthusiasm for walking round the fair looking for new suppliers and original styles. The fair was quieter than normal; many of the Chinese and other foreign exhibitors and visitors hadn't attended. There was a muted atmosphere in the exhibition halls, and I detected an underlying sense of foreboding.

Back in London there was increasing concern about the spread of the virus. On March 9th James Cox, our CEO, left the company. Although the timing wasn't ideal, his departure had long been planned. James had been a successful CEO. He had made Dune an efficient and well-run company. Personally, I had learnt a lot from him. He was excellent at analysing issues in a logical and detailed way. He had a clarity of thought that was refreshing and instructive. James went on to become the CFO of a large food company in Kent near Canterbury where he lived but, like John, remained a non-executive director of the company. We had searched for a new CEO, but none of the candidates were suitable. So, on 9th March 2020, after a gap of 12 years and at the age of 73, I became CEO again. As executive chairman I was deeply involved in the business (some may say too involved). I had had plenty of time for a detailed handover from James which gave a good insight into most areas of the business. We also had an experienced and talented group of directors which made my life much easier. Little did we know that our world would be turned upside down in the coming weeks and months.

After a flurry of instructions about the measures we should be taking in the face of the spread of the pandemic, the British Government finally announced on 23rd March that there would be a national lockdown. All our stores in Britain would be shut, as they would be several times over the course of the pandemic. This posed several unprecedented challenges: apart from how we would manage without store sales, the most urgent was how we would run the company from home.

The response from the team was exceptional. Laptops had been distributed to everyone who needed one. We were all set up on Microsoft Teams and moved seamlessly to virtual meetings. The directors had a meeting on Monday, Wednesday and Friday each week to discuss actions. Having these regular meetings meant that everyone was engaged. Decisions were taken and actions implemented quickly. It was agreed that communicating regularly to the wider team was essential so that they didn't feel isolated. Every Friday I sent out an update on what had been happening that week.

Personally, I wasn't at my best during the first two months of the pandemic. I had contracted Covid around 20th March and felt awful. I had also given it to my wife which didn't go down too well. Fortunately, we didn't have a debilitating cough or breathing difficulties; instead we both had a permanent headache and an overwhelming feeling of weakness and tiredness and took paracetamol every four hours to bring our temperatures down. Everyday chores, like getting in and out of the shower, making the bed and preparing meals were exhausting. At the same time, I was chairing regular online meetings and managing the teams. No doubt the additional pressures of running the business delayed my recovery. I lost a stone in weight. After six weeks we were feeling a lot better although watching the news and hearing the grim statistics, tragic human stories and depressing prospects for ending the pandemic, made us all feel an overarching sense of gloom.

The situation for the business was critical. 30 years of hard work was hanging by a thread. The closure of our stores in the UK, and all our partners' stores around the world, had a devastating effect on the business. We were haemorrhaging money – losing millions upon millions of pounds a year. We had substantial overheads and one of our main sources of income, sales from the stores, had been cut off. Online sales compensated for the loss of store sales to some extent but not enough to plug the gaping hole left by the store closures. The fact that we were predominantly a dressy brand didn't help as slippers and trainers were the footwear of choice during the pandemic. There was limited demand for court shoes, smart sandals and loafers which were a substantial part of our collection. Ironically, the weather was exceptional which, in normal circumstances, would have meant a bumper sandal season.

We needed to speak to suppliers to reduce orders, negotiate a reduction or deferral of our rents with landlords, manage cash and reduce costs. With our shops closed we needed to dramatically reduce our stock commitment. As it was, we had a lot of stock that we couldn't sell. The last thing we needed was more stock flowing in when it was unclear when the lockdown would be over. The message from the government was that the lockdown was likely to last at least a couple of months which would mean missing most of the summer trade.

Our suppliers had their own problems. Every retailer was trying to get out of orders. If the factories accepted all these cancellations, they would be left with a large amount of stock, as well as the components for orders that had been cancelled. We had a small group of suppliers that accounted for over 80 per cent of our purchases. I had known many of these suppliers for over 20 years, like Janna at Maxgreat, Mr and Mrs Kong at King Kong in China and Verno and Wagner Kirsch from GVD in Brazil. We had built a close working and personal relationship with them. We worked as partners and this made a huge difference in agreeing cancellations and discounts. They realised that our very survival depended on their support. Debra and Nilesh did a fantastic job negotiating a deal with them that enabled us to reduce our orders and to get a discount on the stock in transit. We gave a commitment to them to help in any way we could by using components and placing production as soon as the pandemic was over. Their support was incredible and invaluable.

If only the negotiations with the landlords could have been as productive. Unfortunately, that was not the case. Most of our shop leases were for a term of ten years, with anything from one year to seven years left on the lease. They were signed when the market was very different from how it was in 2020. The leases were based on an upwards-only rent, which meant that even though market rents may have come down significantly, as indeed they had, we were still paying a disproportionately high rent. The rent would never come down until the lease came to an end. More recent leases were more flexible and had removed the upwards-only terms, and most had a break clause which enabled the tenant to exit the store after three or five years if trade didn't match expectations. I spent a large amount of my time on the phone to landlords, not only negotiating a temporary reduction in our rent while the shops were closed but also trying to get them to accept a more realistic rent as we were losing a lot of money. The landlords gave us various discounts to help us through the pandemic. As far as a permanent discount was concerned, they were only prepared to consider this if the lease was coming to an end in the next 12 months or so. Unfortunately, this only applied to a few leases. To be fair to the landlords we did have a contract, so they were legally entitled to enforce the terms. However, it was not sustainable for us to continue to pay these unrealistically high rents. We needed a reduction. My pleas fell on deaf ears.

The situation came to a head after the third lockdown on 6th January 2021. We could not carry on losing sales of £1m per week from our stores being closed as it was having a devastating effect on our finances. Reluctantly we decided to use a Company Voluntary Arrangement (CVA), a form of insolvency process that would enable us to compromise payments to landlords and agree temporary new lease terms, while paying our other creditors. We appointed the accountants KPMG to advise us and help us manage the process. Several retailers had used this process to reduce their rents. I wasn't happy about using the CVA, but I had tried everything to get realistic rents from our landlords and had failed. Unless we did the CVA there was a real danger that the company wouldn't survive. The process was only aimed at the landlords so we could have a three-year breathing space when our rents would be at around 10 per cent of sales rather than the much higher figure we were currently paying. A typical example was a store with a turnover of £1m. We were paying rent of £300,000. Under the CVA the rent would reduce to £100,000. If the CVA failed, we would most likely have to go into administration. In that scenario we would lose control of the company. It would be sold by the administrator to the highest bidder. In a distressed sale like this I would most likely write off my investment in Dune. 30-two years of building Dune would be lost. The stakes were high.

To be successful we needed 75 per cent of our creditors to agree to the CVA. Our suppliers and employees almost without exception supported the CVA. Our teams did an amazing job speaking to suppliers, explaining the situation (which for overseas suppliers was not an easy concept to understand). We had a series of meetings with our teams to go through the details of the CVA, answer any queries and hopefully reassure them that this course of action was necessary, and we were confident it would succeed. I was not so successful in persuading the landlords to accept the terms. In the end, almost all of them voted against the CVA as it would set a bad precedent, even though the alternative was a much less attractive option. On 26th February, we were waiting with bated breath for the results of the vote. At one stage the vote was in the balance. It wasn't clear how HMRC would vote. At around 8pm we had a call from KPMG to say that the CVA had been approved by 84 per cent of our creditors. We heaved a huge sigh of relief. Alice Arnold, our Chief Financial Officer, and Clara Eisenberg, our Legal Director, who put in an extraordinary amount of effort, liaising with KPMG and working late into the night, made a massive contribution to bring the process to a successful conclusion. The CVA gave us some breathing space to recover from the effects of the pandemic.

Some positives came out of the pandemic. It gave us an opportunity to step back and reassess our priorities, both personally and as a business. As far as Dune was concerned, we made some key changes to the company during the pandemic that we should have made many years before. We were far too dependent on the department stores, many of which were in terminal decline. In April 2020 Debenhams went into administration. That meant losing a turnover of £16m and a profit of £2m. Fortunately, we avoided losing the commission Debenhams owed us as it paid us 90 per cent of the outstanding amount as we had an agreement to continue trading on condition that they paid our commission.

House of Fraser, which was bought by Mike Ashely's Sports Direct in 2018, was closing stores as it refused to pay the rents that were being asked by landlords. We were one of House of Fraser's largest concessions. Our sales from the concessions, which peaked in 2016 at £50m, had declined by 2021 to £25m. We had a contract to operate the concessions until 2022. However, House of Fraser was keen to move to a wholesale model, as this was their preferred method of trading with suppliers rather than concessions. In May 2021, after protracted negotiations, we closed the concessions and moved to a wholesale model. House of Fraser bought the stock we were holding in the concessions but didn't want to take on our staff. Over a period of a year, we had lost our two biggest concession partners that contributed a substantial part of our profit. On the upside, it accelerated what was a slow and inevitable decline. We had to decide what the business would look like without Debenhams and a different relationship with House of Fraser. Sadly, the closing of the concessions meant that we lost a large team of excellent concession staff, many who had worked for Shoe Studio for a long time. However, it freed up a lot of stock that was sitting in the concessions, much of it moving very slowly. Concessions in many of the smaller towns struggled to attract customers which meant that sales were low, and stock turned slowly. These concessions often only contributed a small profit, or in some cases a loss, which meant that losing them was often a blessing in disguise.

With the loss of the concessions and the ongoing pressures of the pandemic, we had to take action to reduce our costs. We had a smaller business and had to align our costs with our new level of sales. Our biggest overhead was staff costs. We had to make the difficult decision to make several of our head office team redundant, and consequently we lost two of our directors. Given the size of the business, our structure was top heavy. We reduced our head office staff cost by 30 per cent. We made further cuts in 2022 including, sadly, losing three more directors.

One area of the business that did an outstanding job was our distribution centre (DC) based in Leicester run by Roger Crooks and his team. Keeping the DC open and operating was essential to keep our online business going. Despite some of the team contracting the virus and other logistical challenges, the DC continued to operate as normal throughout the pandemic which was an exceptional achievement.

We made the decision to focus our efforts on Dune London and put on hold our sub brands, Head over Heels, Roland Cartier, Roberto Vianni and Bertie. These brands had been used to offer a more varied lower-priced range in the department stores. With the closure of the concessions in Debenhams and House of Fraser, these brands became largely redundant. Losing the House of Fraser concessions meant that external brands, especially men's brands like Boss and Timberland, became a much smaller part of our business as we only sold them in our airport stores and some of the smaller department store concessions. We ended up with a smaller but more sustainable business with our own channels a larger part of our turnover. We were now able to control our own destiny without the reliance on the department stores, although at a significant hit to our profits.

Cash management was essential during the pandemic. One of James Cox's legacies was a strong finance team led by Alice Arnold, our Chief Financial Officer. We had regular and reliable accounts and management information. This proved to be especially important during this period when there were pressures on our bank facilities. It was also critical in our relationship with our bank, HSBC. Being able to accurately forecast our cash requirements, being proactive in providing the bank with detailed information, gave them the reassurance that we were in control. If there is bad news, banks want to hear it well before it happens. Alice and her team had a great relationship with the team at HSBC. They trusted her and believed in the management team. I also had a long and successful relationship with HSBC going back to my Browning days which gave them added confidence in Dune. As a result, we had excellent support from them during what was an exceptionally challenging period for the company.

Style-wise, the pandemic hastened the move to a more casual lifestyle, which meant we have had to change the emphasis of our range from dressy to casual, although our casual shoes still retain a Dune aesthetic. After the end of the last lockdown, in spring/summer 2021, there was a spate of weddings and celebrations that had to be postponed which led to an exceptional season for dressy and occasion shoes. But that was short-lived. Since then, the trend has been firmly to wearing casual and comfortable footwear, in particular trainers, although there is still a demand for smart shoes. However the customer, so used to wearing trainers, wants their dressier shoes to be comfortable as well.

We changed the way we worked with more flexible and hybrid working, mixing working in the office and working from home. I was old school in my approach to working. All my working life I had worked in an office. Working from home was not a concept I was used to. Being the first in the office in the morning at 8am and the last to leave at 6.30pm was a discipline that was ingrained in me over the years. I also dressed for the office. That didn't mean a suit and tie but did mean a shirt and sports jacket or blazer. The pandemic changed all that. I have embraced hybrid working and dress more casually. My wardrobe of suits and jackets has hardly been worn. In some ways hybrid working has made me more productive. If you want to concentrate on a project (like writing this book) there are a lot fewer distractions at home. Having said that, I am still a big advocate of working in the office, especially for young people who need training and mentoring. A minimum of three days in the office is essential to maintain the important exchange of ideas and esprit de corps a company needs to thrive.

There was also a change in people's spending habits that was accelerated by the pandemic. More was being spent on experiences, such as travel and going out to eat, and less on buying stuff, like clothes and shoes. Having had limited opportunities to explore the outside world and socialise with friends and family, the ending of the pandemic restrictions was an opportunity to escape and enjoy those pleasures that had been restricted. There is no doubt that the pandemic had a profound effect on people's mindsets which is still being felt today. It made them more aware of their vulnerability, changed their outlook and led to an increase in mental health issues, especially among the young.

During 2020 and 2021, with the closure of our stores and our partners' stores and the cancellation of wholesale orders from customers, we lost nearly £25m. This was a massive hit to our balance sheet. Over the 30-plus years of Dune's existence, indeed the 50 years I have been in the footwear industry, I had never experienced such a damaging and frightening period. I often felt that we would not get through it. Looking back over the 12 years between 2008 and 2020, the combination of the financial crisis, Brexit and the pandemic posed a set of unprecedented challenges that did immense damage to our business. All of these were external factors completely outside our control. But none of them compared with the existential threat of the pandemic whose repercussions we are still feeling to this day.

The Government's failure to recognise the importance of the retail industry to the economy has added to our challenges. The scrapping of the VAT exemption for foreign visitors has deterred them from buying in the UK, the failure to reform the inequitable rates burden on retailers, the increases in employers' National Insurance and minimum wage, extra red tape and bureaucracy, have all made running a retail business more expensive and difficult.

29. Smaller and sharper

We entered 2022 a smaller and more focused company but with a much-depleted balance sheet. Fortunately, because of the lower rents agreed in the CVA and the post-pandemic spending spree, we made a good profit in the second half of 2022 and 2023, allowing us to partially rebuild our reserves.

After the pandemic we felt that there was an opportunity to open stores in smaller towns around the UK where we had no presence. We had no stores in the Midlands, apart from Birmingham, so we chose Leicester, a city of half a million people in the East Midlands. In East Anglia we didn't have a store, so we opened in Norwich, a cathedral city in Norfolk. Stores in Bromley, Guildford and Cork in Ireland followed. The occupancy costs in these locations were modest so we didn't need a high turnover to make a profit. We felt there was an opportunity for us to become the destination for fashion footwear in these towns. There was also a substantial brand benefit of having a presence in these locations because of the 20 per cent boost we received to local online sales. The visibility of the brand locally raised people's awareness and drove traffic to the website. The problem again was footfall. There just weren't enough customers coming into the stores to generate the sales we needed to make a good return. The message was loud and clear. We needed fewer larger stores in prominent locations in the key shopping centres and high streets as beacons for the brand. The economics of stores in smaller towns was not attractive, a lesson I have been slow to learn.

Measuring the profit of a store has become more complicated. Since the growth of omnichannel trading the store is not just a point of sale. It now performs several new tasks. It processes click and collect transactions (orders placed online and collected from a store). It handles customer returns that were bought online, and it ships orders from the store to the customer in cases where there is no stock of a style in the distribution centre but there is stock in the store. This has put increasing pressure on the store staff at a time when we are looking to reduce costs. We pride ourselves in giving a high level of customer service. With the large increase in staff costs there is a fine line between tightly managing costs against compromising service levels. If you credit a store for all these additional tasks, a store that was marginally profitable could end up making an acceptable profit.

It wasn't all bad news on the store front. By 2024 we had opened 18 outlet stores. These made an excellent return. The outlet centres were typically in a retail park on the outskirts of a town although some were in the country near a major road. Following the US example, outlets had become increasingly popular as they offered well-known brands at 30 per cent off high street prices. Customers often made a special day trip as not only was there a bargain to be had, but there was also a good variety of dining options, easy parking and generally a stress-free shopping environment. We had unsuccessfully tried trading from an outlet centre in Cheshire Oaks Centre near Chester in the north-west about ten years previously. We had made the mistake of offering our sale and end-of-line products. These weren't attractive to customers and the margin we ended up making was too low to make the store pay. We changed our strategy and developed a range of classic styles and styles that were successful but didn't justify being carried forward. The formula worked well, especially once we had enough stores to place good sized orders with the factories which improved our margin. There was an added advantage as in the outlets we paid a turnover rent, which was a percentage of our sales, as opposed to the fixed rent we pay in our full price stores. Many of our outlets now take more money than our full price stores.

The opening of the outlets went hand in hand with our strategy of having "good", "better" and "best" Dune London ranges. The "good" range was made for our off-price business, with the likes of TK Maxx (and its parent company in the US, TJ Maxx), the hugely successful discount chain, where we sold our entry price bags particularly well. The "better" range was built for our outlets and consisted of previous best sellers and high margin styles that we could discount. The "best" range was our main range that we sold in our stores and through our premium partners' distribution channels. It was essential that these ranges didn't overlap as they would have cannibalised our "best" range which was our main brand statement. This was our full price range that we didn't want to discount. With the increased appetite for a bargain there is a danger of taking the easy option and selling lots of the "good" and "better" ranges. Some US brands have grown their outlet business to such an extent that it dwarfs their full price business. Especially when we enter a new market it is essential that our focus is on selling our "best" range, despite the lure of selling lots of diffusion ranges. (Diffusion ranges are secondary collections of merchandise that retail at a lower price, and are often easier to sell.)

After the pandemic, as part of our plan to save money, we cut back on our brand marketing. During the pandemic we had spent a lot of money on Gillian Anderson as the face of the brand. The timing was unfortunate, but I also felt that there was not enough focus on the shoes. We are a footwear and accessory brand. It is essential that the product pings out and is the hero of the campaign. We decided to focus on the product, so the images were of shoes or models' legs wearing the shoes. We didn't have a full model shot which not only made the shoes the hero but also considerably reduced the cost of the photo shoot. The result was a bit dull. An attractive model adds a human dimension and aspirational aspect to the image. The problem is that shoes are always difficult to photograph on a model. If the model is standing, the shoe is only visible at the bottom of the image, where it ceases to be the hero. The eye is drawn to the model's face and clothes before you notice the shoes. One of the skills of a good stylist is to make the clothing neutral so that it blends into the image and doesn't take attention away from the shoes. But that doesn't fully solve the problem. Often the model needs to sit or pose in a way so that the shoe stands out. Sometimes that means the model must be a bit of a contortionist to get in the right pose. Bags are a lot easier as they can be held in many different positions. The combination of a lower spend on advertising and a less interesting creative campaign most probably had a detrimental effect on the brand. Being front of mind as a brand is essential not only to drive traffic to the stores and website but also to raise our profile among the press and key fashion tastemakers.

Our recent creative campaigns have been much more impactful. There is now a greater emphasis on video which gives a greater opportunity to tell a story. Short clips from videos are ideal for social media such as Instagram and TikTok. London landmarks are the backdrop for the campaigns which appeal particularly to international markets and reinforces our brand name, Dune London. We have used the same creative team and models to ensure consistency. In the recent past we had changed the creative too frequently. We got bored and changed our creative direction whereas the customer was only just engaging with the campaign. We needed to resist that temptation. Above all the product is the hero with close-ups of the shoes and bags. We have brought the production in-house with Jamie Brogden, our head of creative, leading the team with input from Debra and myself. This has not only reduced the cost but also ensures the campaign is totally on brand. The ultimate success of a brand's advertising is if you can identify the brand when you cover up the name of the brand on the image. We have hopefully moved in that direction.

We have also started to emphasise our brand codes to build a clear identity for the brand. Apart from the Dune London logo we have the double D (DD logo) and our Dune bug (a decorative trim placed on shoes or as the clasp for bags) which we started using in 2015. We use these on both the products, in our new shop fit, and our different marketing material. Historically we have been guilty of using too many different codes which confused the customer. Keeping with these two brand identifiers and using them consistently has strengthened the brand. We have refitted three of our best stores in our new store design. The emphasis has been in creating a warmer, more elevated environment which has a stronger Dune personality and is an ideal backdrop for our ranges. Both have performed well and encouraged us to refit more of our key stores in this design.

In 2024 we appointed a PR agency, Aisle 8, to raise our profile among the key press and tastemakers. This has included reviving our press days, to showcase the brand and the key stories in our range. These were well attended by fashion editors who were re-introduced to Dune London. They loved the range. We started working with a few strong influencers who really related to the brand and had a good following of potential Dune London customers. We started to get celebrities to wear our hero products and be photographed wearing them. Brand marketing is important for Dune, especially if we want to grow in the US and internationally. The premium attached to being perceived as a "cool" brand is immense. Where we spend our marketing money has changed, but devoting enough money to the channels that have proved successful is essential. How much you should spend on marketing is not an easy decision, as well as how much on performance and how much on brand marketing. The story may be apocryphal but the founder of Skechers, Robert Greenberg, was reported as saying that if he had one dollar left, he would spend it on marketing. If it is true, it has certainly helped make his company very successful.

Women's footwear has always been the dominant product category of Dune, accounting for approximately 60 per cent of our sales. The big change, apart from the move from a formal to a more casual aesthetic, has been the focus on comfort, a trend that is gaining in strength. Consumers are so used to wearing trainers that they are not prepared to accept uncomfortable footwear. We have made huge strides in making our shoes more comfortable by making them softer and more flexible, adding a wide fitting of the key styles and reducing heel heights. There is still a demand for the "wow" high-heeled shoes and sandals as they are flattering, but this has reduced. Now anything goes. You frequently see trainers being worn with a formal outfit.

The other big trend is towards more sustainable footwear. Although fast fashion brands like Shein and Boohoo still have a strong following based on offering fashion at very low prices, there are an increasing number of consumers, especially Gen Z, who are questioning the sustainability of these companies and are demanding a product that is sourced ethically and made using sustainable materials. At Dune all our leather is sourced from tanneries that are approved as gold or silver standard by the Leather Working Group, a body that audits tanneries to ensure that, as well as working conditions, the use of chemicals and water is properly controlled. We also specify recycled materials, where available, for other components used by our suppliers. Often the recycled materials are only marginally more expensive than the non-recycled alternative. The key challenge is changing the manufacturing process, which as I have mentioned previously, hasn't changed greatly over the past 50 years. Using adhesives that are water-based rather than solvent-based is healthier for the environment as they don't pollute water. However, the reality is that 22 billion pairs of shoes go into landfill each year. These shoes are not compostable or biodegradable so will be there for a very long time. Attempts at producing shoes that can be fully recycled are progressing but the types of footwear that meet these requirements is currently limited. Most shoes have some metal in them which makes the cost of recycling prohibitive.

From a product perspective bags have been the biggest growth area of the business. They have gone from 8 per cent of our sales two years ago to 25 per cent of our sales today. Bags and accessories are a lot easier than shoes. For a start there are no sizes, so you need a lot less money tied up in stock. Bags are the key fashion accessory. If you go into any luxury store the first products you see on display are the bags. There are many more successful stores selling bags and accessories (purses, sunglasses, jewellery, watches etc.) than footwear. Kurt Geiger, our main competitor when we had the concessions in the department stores, has successfully reinvented itself from being predominantly a footwear retailer to being an accessories brand. The success of its bags has propelled its stellar growth in the US. Dune relied too heavily for a while on cheaper synthetic bags. We sold them so well there was a reluctance to move to premium, in particular leather, bags. In the last two seasons we have successfully made that transition, and we are selling our premium bags exceptionally well. Over the coming years bags will grow to 50 per cent of our sales.

Men's footwear accounts for 25 per cent of our sales. Competition from other brands is a lot less than in women's footwear. The temporary disappearance of brands like Ted Baker from the high street has opened up a niche for our men's range that is priced between £100 and £150. Our range is particularly strong on formal shoes although with the casualisation of fashion this has dropped from 60 per cent to under 30 per cent of the range. Casuals, trainers and hybrids (a formal upper with a more flexible sole) have become increasingly popular. We have an opportunity to grow our men's sales. Certainly, the product warrants it. Our main issue is that historically we are known as a women's brand and many men don't know we sell footwear. We haven't put enough marketing spend into attracting male customers. This is a big opportunity for the future.

30. Today

Although it was stressful, I had found running the business through the pandemic surprisingly rewarding. I had the opportunity of engaging with the wider management team, gained a better understanding of the broader business and learnt a lot about areas such as IT and e-commerce where my knowledge was relatively sketchy. However, I recognised that I was not the right person to lead the company through the next stage of its growth. We needed a strong CEO with a different perspective.

In September 2022, our long search to find a candidate who was a good cultural fit, had appropriate brand and retail experience, and could deliver on a strong strategic plan, ended. We appointed Nigel Darwin, an experienced manager with plenty of fashion experience. Nigel had had various senior roles in the fashion and beauty sectors. His last position was CEO of the hairdressing brand Toni & Guy. He had a quiet authority and confidence which impressed both the non-executive directors and me. He produced a detailed business plan. One of the key issues he raised with me after a few months in the role was the siloed nature of the senior management team. Teams weren't working closely enough together. This was particularly important given that a lot of the strategic goals required an interdepartmental approach. The structure and personnel of the teams were changed which successfully addressed this issue.

One of the key elements of our strategic plan was elevation. Enhancing both the brand messaging and product quality and display. The new brand creative and celebrity endorsement was being used consistently across all channels. It was raising our profile in the press and with partners. The footwear range was further improved by adding comfort features such as padded socks and more flexible soles, and the handbag range had gone through a major transformation by replacing cheaper synthetic bags with a premium range in leather and special materials. The size of all the ranges was reduced, which made the in-store display more elevated and much clearer to the customer, telling more compelling product stories.

Our biggest growth opportunity was with third-party partners. We had built a successful partnership with John Lewis over 20 years, both in their stores and online, where we were the dominant men's and women's fashion footwear brand. We were also trading exceptionally well on the Next platform where we had seen dynamic growth and where we were one of the most successful women's fashion footwear brands. Marks & Spencer had started selling our footwear and handbag ranges on its website; sales had started slowly but were now showing great momentum. Both Next and Marks & Spencer sold our comfort and wide-fit styles particularly well although, in line with our own experience, customers had a strong appetite for newness. With disposable income being hit by the cost of living crisis, customers needed a reason to buy. Encouragingly our partners were selling our new bag range very well. It was becoming an increasingly important element of the brand.

We were working with hybrid models with our third-party partners which increased flexibility. The main model was consignment, where we shipped the stock to our partner's distribution centre (DC) but still owned it. The introduction of a "drop ship" model allowed us to ship the stock directly to our partners' customer from our DC. This was faster and cheaper and allowed us to offer a wider selection of styles on their websites. This gave our partners the opportunity to trial new styles without the commitment of holding the stock in their DC

With the fall in our department store concession business, our biggest opportunity (and biggest challenge) was growing our international wholesale business, especially in the US and Europe. We had a successful business with Zalando, the dominant online player in Europe, but there was potential to substantially grow this business. In the US, our partnership with both Dillard's and Nordstrom, two of the largest and most successful premium regional department stores, had started well and had exciting growth prospects. We had an excellent range with a clear Dune brand identity. We were building a strong sales team to deliver on this exciting growth opportunity. We had started the transition from being a retailer to a brand, but there was still work to be done.

From 2023 I stepped back from being involved in the day-to-day activities of Dune. It wasn't easy. We had a strong senior management team so I was confident that they would run the business well. However, after so many years, I was used to giving my opinion and being part of the discussion. Now I was partially excluded. It took a few months to get over this feeling. I still went into the office twice a week. I still got all the daily and weekly reports. I am sure I asked too many questions. I tried not to undermine Nigel's position and as time has elapsed, I believe I have largely succeeded. My biggest problem is impatience, maybe because of my age and my character. I want everything done quickly and get frustrated when it isn't. I still attend range reviews and brand marketing meetings. I am still the majority shareholder so I was never going to walk away from the business, but I don't get involved in trading, and my inbox has reduced dramatically.

The past two years have been challenging. The cost of living crisis has made consumers cautious about how they spend their money. Buying fashion shoes (or indeed any shoes except trainers) has not been a priority. Managing costs in today's economic climate is not easy. Many of our costs are increasing above the rate of inflation which puts pressures on margins. In this climate there is a temptation to cut back on areas like marketing. However, there is a danger that this will undermine the investment we are making in the brand. We need to grow our sales. There are exciting opportunities to expand the business and we are seeing these opportunities come to fruition.

When I started Dune in 1992, my vision was for Dune London to be a global footwear and accessories brand. There have been setbacks, many of them outside our control, but we are making strong progress in achieving this goal. I wanted to build a company that was creative and entrepreneurial, that embraced diversity. It was important that the team enjoyed working for Dune; that it was a happy and rewarding place to work. I believe we have achieved this.

I look back on my years in the footwear industry with a great deal of pleasure. I have travelled far and wide and met some exceptional people. It has never been easy. But sometimes managing the difficult challenges is more rewarding and fulfilling. There have been a myriad of obstacles and mistakes along the way. As my father told me, footwear is difficult. He was right. Not just manufacturing but all facets of the industry. Although it may have been less stressful, I would never have enjoyed being an accountant, financier or property developer. Although that could have been an easier way of making a living, I would have been bored. Fashion is exciting and (like a sand dune) is constantly moving. Shoes are a fascinating product which, even after 50 years, I still find absorbing. I count myself lucky to have spent so long doing something that has brought me so much enjoyment. What would my father think? He would be surprised (and no doubt proud) that 135 years after his father arrived penniless in London, and 49 years after joining him manufacturing shoes in his factory in London, his son was still working in the footwear industry.

What I have learnt in business

Over the past 50 years I have learnt many lessons, a lot of them from bitter experience. Here are the key ones.

Hard work

There is no real substitute to working hard if you want to be a successful entrepreneur. My success, in what is a tough industry, is in a large part due to being immersed in my business, putting in that extra effort, and working those extra hours. Many entrepreneurs are obsessive about their business, often to the detriment of their family and social life. They are prepared to make sacrifices in other areas of their life in the single-minded pursuit of growing their business. I personally wouldn't recommend this approach as you need a balanced life. However, I would certainly put my success, in a large part, down to hard work.

Take advice

This sounds obvious, but often we just press ahead without stepping back and seeking advice from someone who has the experience we don't have. It may be a business associate, someone who has had success in your sector, a mentor or friend. Often by discussing issues, you get a clearer idea of the pros and cons of a particular course of action. Never be afraid or too proud to seek a second opinion. In my experience most businesspeople are generous with their time and are happy to give advice. Networking is a good way of exchanging ideas and is valuable in getting a different perspective. Silicon Valley is a great example of how a strong network of tech entrepreneurs, feeding off each other's ideas, has generated amazing success. It is easy as an entrepreneur to become introverted. It is important to have exposure to different ideas.

Do your due diligence

Before you take a job, buy a business, take on a partner or commit to anything that involves the use of your money or time, make sure you have done your homework to satisfy yourself whether it's a good prospect or fit. Once you've made the commitment, it's often not easy to extricate yourself from the situation. We have had this experience with Dune when we have signed up an international partner who has failed to live up to their commitments. If we had investigated the market and the partner in more depth, we most likely would not have made the decision to go with them.

Take a step back and look at the big picture.

Markets change and you need to anticipate these changes and respond to them This is so important. Here I was working very hard as a footwear manufacturer in London in the 1980s. The problem was that manufacturing was rapidly moving to low-cost countries in the Far East. The upside for our business was that it would survive a few more years. The downside (which rapidly became a reality) was that we would have to cease production or fundamentally change the nature of the business. I was so immersed in the detail that I didn't see the big picture and massive changes that were taking place. Anticipating change and seeing the future are major qualities of successful businesspeople.

Be determined and persistent

If your idea is good and, importantly, your execution of the idea is even better, then you will succeed, so don't give up easily, be resilient. In my long career there have been many occasions when I felt like giving up. Customers weren't placing orders, suppliers were late with deliveries, staff were handing in their notice. At moments like those you have to believe in yourself and your business and keep going with added determination.

Fail fast

There is no point being determined and persistent if your business idea or plan is flawed. If things are not working, it is important to fail fast. We have opened stores in locations where it became clear after several months of trading that we would never meet our sales projections. It is better on those occasions to recognise that you have made a mistake. Don't continue in the hope that things will get better. Often, they won't. Of course, you need to make sure that you have not overlooked some obvious factor that is affecting performance. Assuming that is not the case, cut your losses quickly and focus your attention on the profitable and rewarding side of the business.

Don't get distracted

Like many entrepreneurs I have been attracted to new opportunities. The problem is that often means losing focus and taking your eye off your core business. Before you invest a lot of time and money in a new venture make sure that it really is worthwhile and importantly doesn't distract you and your team from the main money-making operation. Sometimes the new opportunity can be game changing for the business. Just be aware of the risks. Make sure you carry out thorough due diligence and prepare a detailed business plan.

Keep control

Once you have sold the majority of the shares of your company, you are not in control of your destiny, which can be demotivating and frustrating. There are occasions when ceding control is inevitable or desirable. If the company buying your shares is going to add a lot of value that you could not achieve without their involvement, then it may be wise to cede control. If it is more about getting the necessary funding, then look at the alternatives to selling shares, like a loan. If you do sell shares get strong professional advice to ensure that the contracts are fair and reasonable and your position as a minority shareholder is protected.

Beware of banks

In my experience, most banks are great when things are going well. It's when things are going badly that you need their support and understanding, and you don't always get it. Because our business requires large stocks, and we have always been growing, we have required large credit facilities. We have therefore been reliant on banks. Overall, I have been very lucky to have supportive bankers. However, we have certainly had some challenging times when banks have reduced facilities or asked for more security, like a personal guarantee. Fortunately, we have struggled through. My strong advice is to avoid giving a personal guarantee. It is also so important to communicate both good and bad news to your banks on a timely basis. Banks tend to be more supportive when you build a close relationship and provide them with timely management information.

Build strong partnerships

Business is very much a partnership with suppliers and customers. Having a strong supply base is essential for a successful business. Over the years, I have been fortunate to build a close relationship with a core of key suppliers and franchise partners, many of whom I have done business with for over ten years. We have supported each other during challenging times because there is a mutual trust and respect.

Build a great team

As an entrepreneur, there is only so much you can do yourself. Many entrepreneurs hold onto the reins for too long and become control freaks. If you want the business to grow, you need to attract the best people and give them responsibility. Be honest with yourself. Recognise your limitations and build a team that can do things better than you.

Learn from your mistakes

I have made a lot of mistakes. Fortunately, the successes have outweighed the failures. It's easy to get downhearted and lose confidence when things go wrong. You have to remain positive. You learn more from your mistakes than your successes.

Love what you do and do what you love

There is no point being an entrepreneur unless you love what you do. It is a tough life that is not suited to everyone. You have to make sacrifices. You won't enjoy it all the time. There will be difficult periods when things go wrong. But if you have a passion for what you do, as I have for shoes, then it is hugely rewarding.

Acknowledgements

Writing *Sole Survivor* has been an enjoyable and rewarding experience. My son, Edward, asked me whether I had kept a journal as he was surprised that I had remembered so much of what happened all those years ago. I didn't keep a journal, and my memory (especially for names and as I get older) isn't that good. However, I now have more time to reflect on the past, and as I started writing this book a whole series of events came flooding back with a clarity that surprised me.

The fact that I ended up in the shoe trade is due to my father, Louis. Although his repeated advice was to avoid footwear, at the age of 28, I joined him making women's fashion shoes and remained in the industry for fifty years. Sadly, he died in 1976, aged 68, so I only spent a year working with him, but that year was crucial in fostering my love of shoes. My mother, Dorie, was always a huge support and encouragement throughout my long career until her death in 2017.

My other mentor in those early years was my uncle (my mother's brother), Len Goodman, who I worked with for ten years. He died last year. He taught me a lot about shoes (he was a very talented manufacturer) but more about loyalty, fairness and how to treat people with respect.

I have always believed that partnerships are an essential aspect of business. My strongest business partner over the past 35 years has been Janna Chen of Maxgreat. She remains Dune's largest supplier. Her work ethic and commitment to continually improving products and processes have been a constant inspiration. I am also grateful to many of our other loyal suppliers, and in particular Mr & Mrs Kong, John Khuu, Debby Law, Verno and Wagner Kirsch, Giuseppe and Christian Di Riccio and Paco Mas for their support over the years, through good times and bad.

Our partnership with Apparel Group has been a major factor in our international growth. I would like to thank Sima and Nilesh Ved for their belief in the brand and establishing Dune London as a major footwear and accessories player in the Middle East.

I had limited experience of retailing when I launched Dune in 1992. I was very lucky to have the advice and guidance of Dennis Fleischer, a consummate retailer, a patient and talented teacher and a lovely human being.

I have been fortunate to have worked with many talented people over the years, both at Browning and Dune, who have taught me a great deal. There are too many to list in full, but I am indebted to the following at Browning: Duncan Miller, Susannah Huller, David Brook and at Dune: John Egan and James Cox (ex-CEOs of Dune and who remain non-executive directors), Zoe Brookes, Barry Marshall, Liz Fenner, Jamie Brogden, Debra Bloom, Nilesh Karia, Alice Arnold and Nigel Darwin.

I would like to make a special mention and thanks to Chris Tanner who was a non-executive director of both Browning and Dune. Apart from being a good friend (whom I met at university), Chris gave me invaluable advice and support over many years, tempering my more irrational proposals with a more sensible and considered approach.

A few relatives and friends read the initial draft and gave valuable feedback. I am grateful to my sister, Naomi Hartnell,

Lili and Brian Massey, Gary Offenbach and Harvey Berk for their comments.

I am indebted to my son, Edward, who edited the first draft of the book. Apart from correcting my erratic spelling he made some important recommendations about changing the order of the book and omitting the overlong sections on my schooling and university life which he felt were unnecessary and self-indulgent.

In researching the book, I had some valuable insights into the manufacturers and retailers in the 1970s and 1980s from Philip Melzer and the early days of Dune from Mohamed Yacoobali. I would like to thank them both.

Martin Hickman of Canbury Press has been an outstanding editor and publisher. He has greatly improved the initial draft by giving it more structure and focus. He has made the whole process a fascinating and enjoyable experience.

My sincere thanks to my children, Edward and Olivia, for their support. They keep me grounded with their constructive criticism and advice. And lastly, I couldn't have got through the trials and tribulations of being in the shoe trade without the love and encouragement of my wife, Anne. Apart from giving valuable feedback on our ranges and being a loyal Dune customer, she has put up with my shoe obsession, the long hours and frequent absences with stoicism and understanding. Family has always been a key inspiration and a fundamental part of our lives. Anne is the one who ensures that these family ties remain strong.

Timeline

1947	Born in London
1960-1966	Stowe School
1966-1969	University of Kent
1969-1974	Touche Ross (chartered accountants)
1974	London & Continental Bankers (bank)
1974-1976	Goodman Price Demolition company
1976-1977	Jack Rose Shoes Limited (women's shoe manufacturer)
1977-1986	London Lane Shoes Limited (women's shoe manufacturer)
1982-1985	Capital Shoes Limited (footwear importers)
!986-2009	Browning Enterprises (footwear importers)
1992	Dune (footwear and accessories retailer)
1993	Open first store at 37A King's Road
1999	Open Browning Hong Kong office
1999	Launch Dune men's range
2001	Open first international store in Riyadh, Saudi Arabia
2005	Dune website launched

2005	First John Lewis concession
2007	Multiple Footwear Retailer of the Year at *Drapers* Award
2009	Dune acquires Shoe Studio and triples in size
2013	Dune rebrands as Dune London
2013	International expansion continues with stores in Philippines and India
2014	Daniel awarded Lifetime Achievement Award at *Drapers* Footwear Awards
2014-16	Multiple Footwear Retailer of the Year three years running
2017	25th anniversary
2018	Open flagship store in Dubai Mall
2018	Open first outlet stores
2022	Daniel made Doctor of the University of Kent
2023	International expansion continues with stores in Saudi Arabia, Nigeria, Australia, Pakistan, Chile and Algeria
2024	Dune London launches in US department stores Dillard's & Nordstrom

Further information: dunelondon.comCanbury by far

Canbury Press
www.canburypress.com

Telling the real story since
2013